Latour–Stengers

For François Gèze, who immediately grasped the importance of the work of Bruno Latour and Isabelle Stengers.

Latour–Stengers

An Entangled Flight

Philippe Pignarre

Translated by Stephen Muecke

polity

First published in French as *Latour–Stengers. Un double vol enchevêtré*. © Editions La Découverte, Paris, 2021

This English edition © Polity Press, 2023

Excerpt from *The Pasteurization of France* by Bruno Latour © Harvard University Press, 1993. Used by permission. All rights reserved

Polity Press
65 Bridge Street
Cambridge CB2 1UR, UK

Polity Press
111 River Street
Hoboken, NJ 07030, USA

All rights reserved. Except for the quotation of short passages for the purpose of criticism and review, no part of this publication may be reproduced, stored in a retrieval system or transmitted, in any form or by any means, electronic, mechanical, photocopying, recording or otherwise, without the prior permission of the publisher.

ISBN-13: 978-1-5095-5550-5 – hardback
ISBN-13: 978-1-5095-5551-2 – paperback

A catalogue record for this book is available from the British Library.

Library of Congress Control Number: 2022945712

Typeset in 10.5 on 12pt Sabon
by Fakenham Prepress Solutions, Fakenham, Norfolk NR21 8NL
Printed and bound in the UK by TJ International Limited

The publisher has used its best endeavours to ensure that the URLs for external websites referred to in this book are correct and active at the time of going to press. However, the publisher has no responsibility for the websites and can make no guarantee that a site will remain live or that the content is or will remain appropriate.

Every effort has been made to trace all copyright holders, but if any have been overlooked the publisher will be pleased to include any necessary credits in any subsequent reprint or edition.

For further information on Polity, visit our website: politybooks.com

Commentary is never faithful. Either there is repetition, which is not commentary, or there is commentary, which is said *differently*. In other words, there is translation and betrayal.[1]

Bruno Latour

Noticing that a situation is entangled calls for disentangling, trying to follow the different threads and separating them ... whereas entangling means lending it more density, greater depth.[2]

Isabelle Stengers

---

[1] Bruno Latour, *The Pasteurization of France*, trans. Alan Sheridan, Cambridge, MA: Harvard University Press, 1988 (1984), p. 178.
[2] Isabelle Stengers, *Activer les possibles*, dialogue avec Frédérique Dolphijn, Noville-sur-Méhaigne: Éditions Esperluètte, 2018, p. 126.

Contents

In Homage to Bruno Latour — ix

Introduction: Speech Impediments — 1

1 To De-Epistemologize . . . — 12
2 . . . Or Disamalgamate the Sciences — 34
3 A Brief Exercise in Empirical Philosophy — 41
4 Sociology or Politics? — 49
5 The *Factish* Gods — 61
6 The Parliament of Things: Doing Ecology — 79
7 Identifying Modes of Existence, Thinking with Whitehead — 97
8 The Intrusion of Gaia — 111
9 Conclusion: Composing a Common World . . . During the Meltdown — 137

Notes — 146
Bibliography — 174
Index — 182

In Homage to Bruno Latour

He loved the world so much . . .

If there is one constant in Bruno Latour's work – which his publishers, La Découverte and Polity, have had the privilege of publishing – it is his love for the world taken as a whole. He neglected nothing, abandoned nothing, eradicated nothing. It was in this sense that he was happy to continue the legacy of the philosopher Alfred North Whitehead.

He worked his way through what he called "modes of existence" (and he identified fifteen of them), delving into them with extensive fieldwork. He loved science. He loved technology (to the point of speaking in the title of one of his books of "the love of technology") at a time when it was fashionable to dismiss it. His brother recently explained to me that when he visited the family vineyard in Beaune, he was interested above all in the smallest details of the wine-making process. His great synthetic tome, *An Inquiry into Modes of Existence*, is also a book that teaches us to love these ways of making the world, despite the way that each often falls into the temptation of mastery.

He shows us how some modes are fragile, threatened with disappearance, as is the case for religion or for politics ("poor politics," as he put it). Both religion and politics are now threatened by something more powerful than themselves: science (but also morality!).

He always wanted to give each mode of existence its dignity, which meant recognizing its own "felicity conditions," its modesty – which, in the end, is its only grandeur. The worst sin is for the mind to confuse modes of existence by reducing them with the help of grand concepts such as the rational/irrational opposition. To judge one mode of existence with the criteria of another is to desiccate the world, to reduce it, to empty it to the point of being unlivable.

That's why he liked activists, as those who learn, and distrusted militants, who know already and only want to convince others. He had therefore launched workshops to collectively explore the world "*in* which we live" and the world "*from* which we live," a way of participating in the environmental movement in its irreducible diversity. To those who reproached him for not appearing to be sufficiently anti-capitalist, he replied that a new class struggle has begun.

Bruno loved "causes," and made Isabelle Stengers' formula his own: "Let the causes cause" [*causer*, to cause, but also 'ramble on']. The paths of these two philosophers are inseparable. You have to read the one to better understand the other. Nor can we understand Bruno Latour without taking an interest in his other fellow thinkers, those of the Centre de Sociologie de l'École des Mines and in particular Michel Callon and Antoine Hennion, but also Philippe Descola, Bruno Karsenti, Tobie Nathan, Donna Haraway, Nastassja Martin, and Nikolaj Schultz.

I learned something from Isabelle Stengers that was very useful when I was with Bruno: if you say you like something – a book, a film, a work of art – you should not stop there. You have to give your reasons. What effect did it have on you, what did you learn or feel? A demanding, sometimes daunting, exercise. If your appreciation wasn't up to the mark, was too offhand or superficial, without attachment, both would soon stop listening to you.

Bruno liked to remind us of the importance of what we are *attached* to. When you come from a world where people swear by emancipation, detachment, and criticism, it is a way of questioning our way of thinking about the world and of doing politics. Why be pushed to talk about deconstruction when all you want to do is give a "good" description of how something achieves its existence?

When I entered the world of research – in pharmaceuticals – before becoming a publisher, I was trying to understand the work of scientists in order to communicate it. At first, I turned to epistemology, but the more I read those philosophers, the less I understood what research work was! That's when Isabelle Stengers urged me to read *Laboratory Life*. Nothing was ever the same again!

Bruno liked the bonds of loyalty. He remained faithful to La Découverte and was one of the reasons why the series "Les Empêcheurs de penser en rond" [Those Who Stop Thought from Turning in Circles] became part of the French publisher's list. He appreciated and always acknowledged the work of Pascale Iltis, Delphine Ribouchon, Caroline Robert (who was careful to produce her books in her own special way), and, of course, François Gèze, who was the first to publish him when he was unknown (and had to be translated from English!). Stéphanie Chevrier, our current manager, was amazed by him.

We saw Bruno enter the last phase of his life, when his country finally recognized him as one of those thinkers "who is the envy of the world" while, with incredible courage, he faced the terrible ordeal of his illness. Our thoughts go out to those who accompanied him throughout, and in particular to his children and his wife, Chantal.

He leaves us with one question: how will we carry on his legacy?

Philippe Pignarre

Introduction: Speech Impediments

I conceived of this book as a kind of patchwork composed of many quotations, which might give the reader a somewhat unstable feeling. But I thought that engineering it in this particular manner was the best way to come to terms with the comings and goings between the works of Bruno Latour and Isabelle Stengers, a particular mode of "weaving"[1] which made me adopt as a title Gilles Deleuze's admirable phrase, quoted by Stengers, "an entangled double flight."[2] It will thus run a zigzag course. Citing them both at length meant I chose to dramatize in a certain way because simply summarizing their different texts would not have worked. I wanted the reader to be touched by their actual modes of expression, being as close as possible to them, while my undertaking of an overly pedagogical task of exegesis would deprive us of their brilliant flashes of thoughts often grasped in full flight.

In the first place, I took this work on as an editor who loves the authors he is publishing. After all, what is an editor, in the end, if not the first mouthpiece for the texts he has chosen to uphold? I hope this will generate the desire to delve into the respective works of Latour and Stengers that I hold so dear, and of which I have no hesitation in saying that they have changed in the deepest way my manner of being in the world. I wanted to sharpen your appetite. Reading Latour's *oeuvre* by regularly confronting it with Stengers' propositions – was this the right attitude for plunging into such a witch's cauldron? Each of you will make up your own mind. Latour and Stengers have descendants in common who know their works well and will put them to good use. But nonetheless I think I was the only one able to sit down to this task and who had the time, "profiting" from the isolations of 2020 and 2021.[3]

The procedure I follow here is not entirely symmetrical. I have not tried to paint a picture of how Stengers' thought was built up, from her meeting with Ilya Prigogine to reactivating the work of Alfred North Whitehead, passing by way of the intellectual encounters she had with Léon Chertok or Tobie Nathan. Attempting a parallel task with the *oeuvre* of Bruno Latour would be mission impossible. So I voluntarily chose to concentrate on the latter, privileging as much as possible the points at which it crosses, collides, or aligns itself with Stengers, who quite often quickly seizes on Latour's propositions but without ever leaving them intact. Rather, she makes them twist in a way that one could call political. I pay particular attention to those moments where Stengers puts in her own words what she has learned (or taken) from Latour and also to the ways in which she emphasizes both their importance and her divergences. As for Latour, when he comes to meet Stengers, it is often through a shift in his arguments – his own ones; so that has to be inscribed in the movement of his ideas without suddenly bursting in. As it happens, in the course of time, the references to Stengers multiply in his writings to the point where he dedicates *Politiques de la nature* to her in 1999.[4] (Stengers had dedicated *L'Invention des sciences modernes* to him six years earlier, "For Félix Guattari and Bruno Latour, in memory of a meeting that did not take place."[5]) This is how Stengers speaks of her relation to Latour: "[His] subtle and demanding reading [of the first draft of her book, *Au temps des catastrophes*] is written into the process, which for over twenty years is witness to the fact that agreements among sometimes divergent paths are made thanks to the divergence and not in spite of it."[6] There are also notable occasions when one of them says how and why they are borrowing a proposition from the other or how it is to be understood.

If this book follows a chronological path, it is nonetheless replete with references which are often not in that order. It seemed to me useful to put in formulations that were able to throw light on propositions made earlier, but which one of the authors had fully explained, often with different words, only at a later stage. And again, in that sense, this book is "woven."

One of the difficulties of this task comes from the differences between their respective styles. Like two magnets, Latour and Stengers are attracted and fascinated by each other's conceptual propositions, but they are quite distinct in their ways of writing. In order to be convincing, Latour multiplies his pedagogical exercises and is happy to be repetitive, to demonstrate again and again, in order to make them more accessible. He creates characters (like the young anthropologist who questions him in *Cogitamus*, and who turns up again in his *Enquête sur les modes d'existence*).[7] He multiplies conceptual inventions and even shock formulas (Irreduction, Moderns, Great Divide, black boxes, factishes, Parliament of Things, Double Click, to de-economize . . .), examples, explanations in boxes, diagrams, paintings, drawings, extracts from comics, photos, and theatrical set designs. The disorder in Latour's multiple interventions and ways of intervening is only apparent. He often says that, because of the irruption of Gaia, one has to use everything in one's arsenal, for how else can we find forms modified to the representation of this new cosmos that is nonetheless ours?

For her part, Stengers is quick on the uptake as she multiplies her propositions (requirements and obligations, speculative thought in the strong sense, the cosmopolitical, diplomats, slowing down, recalcitrance, modes of abstraction, induction . . .). One should not miss a single sentence in her argumentation because the occasion to catch up later will not occur. One has to understand straightaway, and so be prepared to slow down as one reads, or go back over her text. Reading her books is not a frolic in the woods. You have to read the chapters in order. Her thought is tight, precise, and moves forwards implacably. But do not think that Stengers writes without hesitations. If you have access to the different versions that have emerged successively from her pen, you know that that is far from the case. Latour has turned himself into a sociologist, ethnographer, historian, philosopher, but always a researcher. As for Stengers, she is a philosopher, irremediably a philosopher, including in the two fictional works she has written, the first on Freud and the second on Newton and Leibniz.[8]

Yet there is something they do have in common, something a little obscured, or at least difficult to grasp because it relates to a philosophical question that will turn out to be of prime importance. Anyone who has attended a public occasion on which Stengers is speaking will have been struck by her hesitations, with her sentences interrupted by a "How should I put this?" which may not just be anecdotal. What kind of Latouro-Stengerian interpretation could one give of this? One would be mistaken, of course, to imagine the hesitation has any kind of psychological basis when in fact it is a matter of the problem to be solved, with the proposition itself in the process of bursting forth, asserting itself in the murmur of the world, something that is difficult to express with precision. It is indeed the need to "depersonalize the experience of the work-in-process, that is, get rid of anything that gives it a psychological or social narrative."[9] Everything that needs to be said is still virtual. This is much more like the hesitation of a mountaineer on a difficult alpine climb, looking for the best grip on a vertical wall. How does one get a grip? Adopting the point of view of the climber is not enough because there is the mountain as well – or the audience, for our case at hand. But it is perhaps the example of the surfer that is the most eloquent:

> with each wave, surfers take the risk of catching it or letting it go; they have no illusions of being in control. What is at risk is their possibility of keeping on, of sliding into the wave, at the critical point where only a precise and sensitive insertion of one motion into the other can make them earn the respect of the breaking wave.[10]

Didier Debaise will put it like this: "You can't just decide that you have a soul, an idea, or a feeling: they grab you from the outside."[11]

Stengers' "How should I put this?" is the equivalent of Latour's paintings, graphics, and diagrams that punctuate his books. He has offered a very nice formulation to describe these instances of "How should I put this?" as *speech impediments* that designate "not speech itself but the difficulties one has in speaking and the devices one needs for the articulation of the common world – to avoid taking *logocentric* words . . . as facile expressions of meanings that would not need any particular mediation to manifest themselves transparently."[12] He goes on to

Introduction: Speech Impediments 5

elaborate that "the connotations of the word [*articulation*] cover the range of meanings that I am attempting to bring together, meanings that no longer stress the distinction between the world and what is said about it, but rather the ways in which the world *is loaded* into discourse."[13] He will open his *Rejoicing* book in the same way.

> Rejoicing – or the torments of religious speech: that is what he [Latour] wants to talk about, that is what he can't actually seem to talk about: it is as though the cat has got his tongue; as though words were impediments; as though it was impossible to articulate; he can't actually seem to share what, for so long, he has held so dear to his heart . . . he can only stutter . . .[14]

As it happens, Stengers also turned "the idea that flees if one tries to make it explicit"[15] into a philosophical question in her *Thinking with Whitehead* book:

> The point it neither to describe nor to explain but to produce a set of constraints that impose on thought a regime of reciprocal presupposition. A "leap of the imagination" may respond to these categories, but it is a vertical leap, conferring on words the capacity to evoke, not to designate. It is not that process "transcends" language, but what is appropriate to it is the component of stammering in language, the "Well, what I mean is . . ." or the "How should I say . . ." in which what hesitates is not a set of potential statements but the very wording of the words, together with the "I" who "means" [*veut dire*].[16]

This picks up on William James's "undecidable question: am I touched [*ému*] because the world is touching [*émouvant*] or does the world seem to me to be touching because I am touched?"[17] It happens that *something* emerges as an argument proceeds, wending its way and interrupting "the automatic interpretation that makes me attribute either to an external cause or to a reason of mine the fact that an experience has passed . . .";[18] it happens that a proposition is difficult to formulate because it does not relate to some solitary cerebral exercise on the part of the speaker but presupposes a *leap of the imagination* in order to be formulated with all the hesitations, the risks of betrayal, that are part of its other engagement, this time with an audience that has to be up to the task of listening, sharing the hesitations, sensing that this work-in-progress might fail. "Even the wise Plotinus, reaching towards Intellect, must have a discursive practice based on the experience that he has, which is fragmented, problematic, discordant, an approach in which resides the tension towards the unity of creating/producing/discovering, together with the intense awareness that he could be mistaken."[19] We shall see that here lies one of the many facets of what Latour will quickly come to call the Great Divide, cropping up in all his work.

For Latour, a "musical metaphor" is a good way to come to terms with this situation: one can hear a *melody* that remains inaudible for those not involved – "a melody to which we become better and better attuned." It is not a question of saying we have a "mind zooming toward a fixed – but inaccessible – target. It is the fact that 'occurs,' that emerges, and that, so to speak, offers you a (partially) new mind."[20]

Stengers will even propose the term "induction" (in a different sense from the traditional meaning), a word she learned with Chertok, and which refers to the relation between the hypnotizer (or hypnotist) and the hypnotized, to qualify a fairly unique creative situation: when the idea "flees if one tries to make it explicit," when the "enigma puts the creator to the question," when the creators are "creatures of their question,"[21] when propositions "possess individuals far more than individuals possess them,"[22] when she speaks of the experience "of those who know that what they seem to be the authors of is in fact what obligates them."[23]

Because "hypnotizers are well aware that they are not the ones who have given the order [for example, asking the hypnotized person to raise their arm]. If they have a role, it is rather that of indicating a path, or authorizing an experience."[24] She will return to this with a more technical account in 2015:

> The achievement of saying to oneself "I understand" is not an act of thought. The "understanding", as much as the "I" who has understood, both owe their existence to a path of instauration, a response appearing in the wake of "something to be understood," a double and correlative grasp by the form of both the agent of the instauration and the thing instaured.[25]

This is also what this book is trying to do by "tracking" Latour and Stengers.[26] It attempts what Stengers calls a "speculative gesture," rather than a boring pedagogical exercise.²⁷ Making a speculative gesture means deploying the experience in all its dimensions, including with virtual ones that accompany it without becoming apparent.

Stengers herself makes the link between Whitehead and another philosopher, Étienne Souriau, whom we shall see will also be all-important for Latour as the two philosophers contribute, each in his own way, to break the spell cast by "the subject facing the object," epistemological abstraction. Here is how Stengers is citing Souriau in *Thinking with Whitehead*:

> I insist on this idea that as long as the work is in the workshop, the work is in danger. At each moment, each one of the artist's actions, or rather *from* each of the artist's actions, it may live or die. The agile choreography of an improviser, noticing and resolving in the same instant the problems raised for him by this hurried advance of the work . . . [or] the works of the composer or the writer at their table . . . all must ceaselessly answer, in a slow or rapid progression, the questions of the sphinx – guess, or you will be devoured. But it is the work that flourishes or disappears, it is it that progresses or is devoured.²⁸

Latour puts it like this:

> To say, for example, that a fact is "constructed" is inevitably (and they paid me good money to know this) to designate the knowing subject as the origin of the vector, as in the image of God the potter. But the opposite move, of saying of a work of art that it results from an instauration, is to get oneself ready to see the potter as the one who welcomes, gathers, prepares, explores, and invents the form of the work, just as one discovers or "invents" a treasure.²⁹

I should immediately warn the reader, who may get a surprise, or even be disturbed, when they encounter this somewhat awkward development: it is no minor matter and we shall see it emerge once again in chapter 7 of this book under the heading of the "bifurcation of nature," when we shall also meet Souriau again, and Whitehead, of course. In a book published in 2020, *Réactiver le sens commun: Lecture de Whitehead en temps de débâcle*, Stengers will introduce the idea of the "middle voice," contrasting with, on the one hand, the active voice where the syntactic subject designates the entity acting, and the passive voice where the syntactic subject is the one undergoing the action.[30] She writes:

> Bruno Latour, however, contributed somewhat to the resuscitation of the semantic pertinence of the middle voice by proposing that we hear it in instances in which we hesitate over the attribution of an action . . . Instead of associating the middle voice with a general acknowledgment that we are not the sovereign authors of our actions, Latour proposes to associate it with concern and care over our manners of being attached.[31]

This is clearly one way of thinking that Latour and Stengers share.[32]

But to return to the question of their different styles, is it unimportant? It would be too simple to say yes. In one of his first writings, on Péguy, Latour made a point of emphasizing what can be learned from style. On Péguy's notable tendency for repetition, he wrote: 'This simple impression stops us from seeing this aspect of his style as a problem of *form* . . . for the moment we have to consider Péguy's repetitive style as the *basic* problem of his work . . . formal effects need to gather up the movement that basic content can only capture as a betrayal."[33] A seeming paradox: things must be *touched on lightly* in order to better *grasp* them, an idea appearing again in the *Inquiry into the Modes of Existence* book, first published in French in 2012, which brings to a close numerous inquiries, but which also marks, as we shall see, the definitive refusal of any discourse of *critique* that purports to *unmask*, to show what there is *behind* things.[34] What is true for Péguy may also be true for our two authors. We shall see the importance of *reprise*, *starting over again*, being reactivated at the moment when Latour and Stengers reread and jointly present Souriau, or when Stengers writes *Thinking with Whitehead*.

Another difficulty comes from the fact that they don't make a point of writing for each other. They have plenty of other interlocutors (in particular, Algirdas Greimas, Harold Garfinkel, and Françoise Bastide in Latour's earlier works, but also Philippe Descola or Michel Callon right through his career). Stengers likes to recount that her "first native habitat was the novels of Alexandre Dumas":

Introduction: Speech Impediments

When I say that I can sometimes hear the echo of Dumas in what I write, it is perhaps because I liked his powerful characters. All his characters are strong, and he liked them all for their strength and their clashes. That is what I want to do when I write. I want my protagonists to speak with all their force, even if it is a dark, ominous force. . . . And I'm hoping that that will nourish readers' thinking forces.[35]

And then, Latour and Stengers don't do their politics in the same way. They are not looking for the same allies. Hence Latour has often exasperated the Marxists (at least in France, where for convenience they make him out to be a relativist), while what Stengers has to say has apparently been more acceptable to them. We shall see that it is a bit more complicated than that. But, in both cases, the Marxists have often taken the precaution of avoiding them and not reading them, despising them rather than confronting them. Stengers has increased her contacts with activists of all stripes, from electromagnetic hypersensitives to those ripping up GM plants, passing by way of the Zadists as well. She had a translation done into French of the neo-pagan witch Starhawk, having announced her deep-felt regard for her. She leapt to the defense of Houria Bouteldja, one of the founders of the *Indigènes de la République* party. Latour, more often, has kept away from direct engagement.

Latour and Stengers share a good deal of common philosophical background: first, the American pragmatism of William James and John Dewey,[36] then Alfred North Whitehead, Gilles Deleuze, or, closer to us, Donna Haraway. They also began their work on the sciences in the company of Michel Serres, even if they parted ways with him later.

But another thing that their two *oeuvres* have in common is their amazing continuity. Readers with a tendency to begin with the latest works off the shelf would be convinced of this by delving into *Irreductions*, first published by Latour in 1984 as an appendix to his book on Pasteur. Or, Stengers' *The Invention of Modern Science*, first published in French in 1993. They don't look a day older. Of course, the words used are changed, refined, renewed. Words are worn out and get older, sometimes because others take them away to say something else with them, rob them of their effect, or even destroy them. But such work also allows for precision to be applied, or for finding new, more adequate, formulations, even to the point of inventing terms or borrowing them from another language. If the words change, it is also because of a difficulty that Latour, rather pessimistically, brings to the fore: "People have ideas on the sciences that Voltaire had in the eighteenth century, and they have not budged one centimeter."[37] Latour humorously noted, as early as 1999:

> The science wars, from this standpoint, are not lacking in a certain grandeur. I would join the camp of the "Sokalists" [allusion to the Sokal and Bricmont affair, two scientists behind a hoax designed to provoke a violent denunciation of the human sciences getting involved in the experimental sciences] right away if I heard someone calmly proclaim that the sciences are one "system of beliefs" among others, a "social construction" without any particular validity, an interplay of political interests in which the strongest wins (positions that are usually attributed to me by people who have not read my work!). "*That means war!*" as Isabelle Stengers reminds us.[38]

And Stengers confirms:

> They still always point out to him [Latour] that if he maintains that neither reason nor nature has the last word in this production [of scientific objects], then the objects are nothing more than relative, like everything else, to arbitrary opinion and blind political relations. A rock-solid protestation, obstinate and deaf to any argument. It is well and truly an order word that Latour has come up against.[39]

This book could also have been entitled "Les causes de Bruno Latour et Isabelle Stengers," relating to the double meaning of the word *cause* (what makes us think and what must be left to happen): "It is simply that the notion of cause is not sufficient in itself, not any more than the notion of explanation. One could say that each cause poses the question of how, here and now, it is going to cause."[40] Many years later, opening a Cerisy colloquium, Stengers adds: "Thinking is not 'thinking on' or 'thinking about' but 'because of' [*à cause*]."[41]

If this little book wins its bet, it will be that the reader will have understood that there are not two Latours, the first a sociologist of the sciences (being more or less relativist), and a second who suddenly got interested in ecology (heaping praise on scientific institutions and in particular the IPCC, the intergovernmental body of the United Nations responsible for knowledge on climate change), this latter interest having nothing much to do with the former. We shall see that it is thanks to his work on the sciences that he was able to think ecology afresh, to completely redefine it. And he understood how important Stengers' dogged work was, for while it regularly connected with the Latourian ideas, it kept unfailingly to its original philosophical path. It is an opportunity to work with both, following what Donna Haraway has called their "companionable friction."[42]

I have obviously not written this book without trepidation. Will I be up to it? Am I not presumptuous? And what if one of these authors, or even both, thinks they are badly treated, misunderstood, underestimated? So it is with these added pressures that I have embarked upon this adventure.

1
To De-Epistemologize...

I begin at a micro level: two innocent-looking footnotes that transport us from the study of science as it is carried out in laboratories to the awareness of Gaia. The first is from Latour's *La Science en action*, a book that you might recall was first published in English in 1987 before being translated into French[1]: "One can read with interest Ginzburg's (1980) counterexample which, he thinks, can separate the *sciences* of the trace or symptom from 'exact' sciences."[2] The second is from Stengers, in *The Invention of Modern Science*, first published in 1993:[3] "On this subject, see how Carlo Ginzburg contrasts the sciences of proof and the sciences of indices."[4] The same reference, but on the one hand a doubt ("He thinks it can") and on the other an affirmation ("On this subject, see . . ."). Is Ginzburg's suggestion concerning the different ways of doing science merely superficial, or will it be included, in various ways, in subsequent debates between our two authors? What will they make of it in the course of their respective works?

Ginzburg is an historian, and also a philosopher. The example he is working on here is about the dating of a picture in art history. In order to distinguish a real painting from a fake, the expert has to focus on "the examination of the smallest details where the influence of the school to which the painter belongs is the least marked – which is the case for the earlobes, the fingernails, the shape of the fingers and toes." "The art connoisseur and the detective may well be compared, each discovering, from clues unnoticed by others, the author in one case of a crime, in the other of a painting."[5] He concludes:

> This "low intuition" is rooted in the senses (though it goes beyond them) – and as such it has nothing to do with the extrasensory intuition of various nineteenth- and twentieth-century irrationalisms. It exists everywhere in the world, without geographic, historical, ethnic, gender, or class exception; and this means that it is very different from any form of "superior" knowledge, which is always restricted to an elite. It was the heritage of the Bengalis whom Sir William Herschel expropriated; of hunters; of mariners; of women. It forms a real link between the human animal and other animal species.[6]

Ginzburg is thus suggesting a way to exit from what he calls the dead end between rationality and irrationality by putting the accent on a practice that does not relate to the modern experimental sciences, but nonetheless deserves to be identified and celebrated. Now we understand a little better why Latour and Stengers read him carefully.

It would obviously be absurd to want to summarize or pin down, to just these two footnotes, the incredible intellectual exchange that has gone on for several decades between these two authors, these two friends. Yet they can serve as a marker for the way in which Latour's work has been discussed by Stengers, and opened up a space for a constant coming and going which has nourished a large number of authors working on science and technology, and especially today those engaged in struggles around the "New Climatic Regime" – which Stengers calls our "catastrophic times."

If this common reference – hesitant in one case, firm in the other – appearing very early in their respective works, is so interesting, it is because it alerts us to sciences that are not defined as "experimental," that are not based in laboratories in the strict sense (except when one broadens the definition of laboratory, which Latour will do sometimes). It turns out that they will also be, for the most part, the very sciences – like geology, ethology, or climatology – without which we would be at a disadvantage when it comes to facing Gaia. But let us not jump ahead to our conclusion!

How, then, can the sciences be put in their place? How can we prevent "Science" (with a capital S) cannibalizing all knowledges, all practices? If both Latour and Stengers are ambitious about answering these questions, they will nonetheless go about it in very different ways, although through a common operator: they are going to embark on disarticulating Science without going by way of simple "critique." Neither of them is looking for what drives their actors – scientists in this case – without their knowledge. They do not want to *demask*; rather they want to *characterize*, by striving better to "get a fix" on the actors and onto their work, to speak better about what they are doing and thus to render the old epistemology, in particular the Bachelardian one, obsolete. That epistemology could be called "critical" in that it thought it was revealing the background to the scientists' work, in which the idea of Reason was central. And for both Latour and Stengers, it was indeed a matter of going back to scientific *practices*. For Latour, "'practice' is not something that you observe *de visu*, but it is more a method of observation. It is a genre that may retrieve as much from dead documents and immensely distant times as from visible sites."[7] And, for Stengers, "the practices of a scientist, a technician, or a lawyer imply a particular art of attention; these practices permit them, even demand of them, when they are not automatic, to hesitate and to learn."[8] Latour echoes this when he writes that it is a question of proposing "concepts that try to capture the actual experience of multiple actors, in terms that put them less in contradiction with themselves."[9]

Now I shall dwell at some length on Latour's work, before coming back to that of Stengers.

Could there be anything else that makes us prouder of who we are, we who Latour calls the "Moderns" (Latour has inherited from Charles Péguy the idea that the "modern world" was a "universal disaster," a "monstrous disruption" born of the aborted social revolution[10]), than our special relationship with the sciences? Probably nothing, which is why we had to begin with them to carry out our auto-ethnography, just as we have done the anthropology of others. Everything began in 1979 with the publication in English of *Laboratory Life*, written with Steve Woolgar. But this work on the anthropology of the sciences will be taken up again in multiple ways, eventually stretching over more than forty years.[11] So it is quite reasonable to think that the science studies initiated in the laboratories of Professor Guillemin – who will receive the Nobel Prize in Medicine – at the Salk Institute in San Diego, California, was, from the very start, but a detour in a larger project: understanding ourselves, we who attribute so much importance to the sciences we are inventing and mastering. This ambition is also in evidence in the other inquiries carried out by Latour, on religious speech, or the fabrication of the law,[12] but it is clearly expressed in *Irreductions* from 1984.[13] This essay is quite amazing in that it precedes most of Latour's ethnographic studies, and can be closely aligned with his later philosophical ideas, in particular those that will appear in *Facing Gaia* and the two following books, *Down to Earth* and *After Lockdown*. He himself refers to it as his "Tractatus Scientifico-Politicus," alluding to Spinoza's *Tractatus Theologico-Politicus*.

From the start, Latour chose not to take "capitalism" as his point of departure because he thought it too narrow ("Capitalism is still marginal even today. Soon people will realize that it is universal only in the imagination of its enemies and advocates"[14]); he would look to the larger picture of the "modern world." In order to do this, he borrowed from Michel Tournier's *Friday, Or the Other Island*, a rewrite of the Robinson Crusoe story. When Friday, in this novel, accidently blows up the gunpowder carefully recovered by Robinson, the latter finds he is "naked as he was on his first day." So he begins to follow Friday who, he discovers "lives on an entirely different island . . . while Friday finds himself among rivals, allies, traitors, friends, confidants, a whole mass of brothers and chums, of whom only one carries the name of man."[15]

Robinson lived in a well-ordered world, a world he thought was simple and *in which he was alone*. What he discovers is a mixed up, *implex*, world, more inhabited, complicated, and rich. "How many actants are there? This cannot be determined until they have been measured against each other."[16] What is the force of an actant (this term can be used for nonhumans, keeping the term "actor" for humans)? It is its capacity to associate with other actants. So they have to be followed patiently in all their twists and turns, without grouping them into the two categories dictated by the Great Divide, *subject and object*, or *society and nature*.

The Moderns' power consists in associating – under the cloak, so to speak – forces that appear irreducible the ones to the others: science, religion, technique, geography, economy, the army . . . "They [the Moderns] believed in a separate order from which they drew their strengths."[17] They gained their strength by coming together, "each one separated and isolated in his virtue, but all supported by the whole. With this infinitely fragile spider's web, they paralyzed all the other worlds, ensnared all the islands and singularities, and suffocated all their networks and fabrics. Those who 'invented the modern world' were not the strongest or the most correct, and neither are they today."[18] Stengers also announces: "We who have become modern and have learned not to be slowed down any more by poisonous concoctions, we are the ones who have become capable of carrying peace and civilization elsewhere."[19] Which is to say the destruction of others . . . She will later return, in 2002, in her first book on Whitehead, to the link suggested by Latour "between incoherence and destructive power": "the interminable quarrel is precisely what makes of modernity an invincible war machine." How?

> This machine's discordant unanimity lies in the derision it reserves for those designated as "still imprisoned in their beliefs." The verdict converges, in their case, whether coming from missionaries or scientists, Hegelians or the postmodernists who deride the Hegelian grand narrative, preferring to have charming conversations. This verdict is "they are going to have to learn to take part in our common destiny." The chips are down, nothing will escape our quarrels, not one earthly inhabitant, not one human practice.[20]

But in the end Robinson is not all that different from Friday, and will go the opposite way from the Moderns. He too will happily mix up different actants, but with the one difference from Friday: *he absolutely denies it.* "So, you were wrong, Crusoe. There is no modern world to be set against your primitive island."[21] The world he was living in and feeling so alone in (alone with his objects – even Friday is reduced to slavery) was the fruit of lies and hypocrisy. "We always make the same mistake. We distinguish between the barbarous and the civilized, the constructed and the dissolved, the ordered and the disordered."[22]

Hence the urgency to do an anthropology of the Moderns using the tools forged to do just that with Others. Latour writes that, in 2012, after recalling having begun his career on a fieldwork mission to the Ivory Coast, he was struck by:

> a flagrant asymmetry here: the Whites anthropologized the Blacks . . . but they avoided anthropologizing themselves. Or else they did so in a falsely distant, "exotic" fashion by focusing on the most archaic aspects of their own society . . . and not on what I was seeing with my own eyes . . . industrial technologies, economization, "development," scientific reasoning, and so on.[23]

We have to understand where the Moderns were drawing their power from, this power that enabled them to annihilate the others. It is not because Latour speaks of the Moderns that we have to believe that we were never modern. And if we are, it is not for the reasons generally given. It is presenting ourselves as Moderns, while not being such, which gives us our particular potency: "For years we have *voluntarily* granted to the 'modern world' a potency that it does not have. Perhaps once upon a time it bluffed and claimed superiority, but there was *no reason whatsoever* to concede this superiority."[24]

A general lesson can be drawn from *Irreductions*. By allowing oneself to believe what the Moderns say about themselves, differentiating them from all the others, while also allowing oneself to believe what capitalism says about itself, we condemn ourselves to starting up a war machine, a destructive potency that is impossible to control.

> What we are pleased to call "other cultures" have a number of secrets; ours only has one. This is why "other cultures" seem mysterious to us and worth knowing, whereas our own seems both unknowable and stripped of mystery. This secret is the *only* thing that distinguishes our culture from the others: that it and it alone is not one *culture* among many. Our belief in the modern world arises from this denial alone.[25]

The modern world rests, then, on a "Great Divide," for which Latour gave a definition we could call "decolonial" as early as 1983:

> The expression, "the Great Divide" has allowed a lot of authors to sum up the division they thought they were observing between the scientific and the prescientific mind, a division that also cuts off

modern occidental societies from "other" societies. . . . The project [Latour's] . . . was to believe nothing a priori about the Great Divide, and to submit all knowledge productions to the same kind of ethnographic inquiry, whether Alladian sorcerers, American biochemists, Greek mythologists, bush intellectuals, electrical engineers, New Guinean botanists or the Queen's botanists from Kew Gardens . . . It is thus that many of us, at the same moment, passed from prescientific fields to scientific ones and set about studying the experts, carrying away with them the questions, methods, and research procedures appropriate for ethnography.[26]

When he wrote *Science in Action*, Latour followed up this work begun five years earlier, and he could have given the book the title of the fourth chapter of *Irreductions*, "Irreductions of 'the Sciences.'" He takes up again the question of the sciences with the help of the following principle, which is expressed through the challenge of the Great Divide: radical separation of Nature and Society; facts and values; but first, here, between science and opinion – or common sense – and between those who "know" and those who "believe." This separation structures all epistemology, a philosophical discipline which has the objective of grasping what defines science, what puts it in radical opposition with other modes of thought thus disqualified. Here, for example, it is presented in a way that is too good to be true, in this citation from Étienne Balibar reporting a dialogue between Alain Badiou and Georges Canguilhem:

> Q *(AB)*: Should one continue to radically oppose scientific knowledge and common [*vulgaire*] knowledge?
> A *(GC)*: Yes, and increasingly. There is no scientific knowledge without, on the one hand, very elaborate mathematical theories, and without, on the other hand, the manipulation of instruments of greater and greater complexity. I'm quite happy to add that there is no common knowledge.
> Q: Do you understand the expression "scientific knowledge" to be a pleonasm?
> A: You understand me perfectly. That's what I mean. Knowledge that is not scientific is not knowledge. I would maintain that "true knowledge" is a pleonasm; ditto "scientific knowledge" and "science and truth." They are all the same thing. That does not mean that there are not, for the human mind, any aims or values that lie outside of truth, but that means that you cannot call knowledge what is not knowledge, and that you cannot give

this name to some way of living that has nothing to do with the truth, that is to say with rigor.[27]

The aim of getting rid of the Great Divide – and all that flows from it – is one of the common threads running through Latour's work, as also, in sometimes different ways, in that of Stengers. In order to get there, Latour invented a sociology, or an anthropology, even an ethnography, of the sciences, in which the principal tool is the most detailed possible description of the work of scientists, what came to be known in the English-speaking world as science and technology studies (STS).[28] So, to understand the sciences, one has to forget about epistemology and *start afresh*, and, to do that, head off to the laboratories to find out how the *researchers* experiment, write their articles for the science journals, have arguments with their colleagues. This sociology or anthropology should make the *materiality* of what the researchers do visible, which they *don't always see the point of talking about*.[29] Stengers writes in the same perspective: "The manner in which scientists present the sciences, what they identify with science, the consensual reasons to which they have recourse, the arguments that allow them to denounce the 'rising tide of irrationality . . .,' all of that is literally saturated with borrowings from philosophers."[30] According to Latour, "Perhaps epistemology is a confusion of the senses. We follow the dazzled gaze but forget the hands that write, combine things, and mount experiments."[31] He goes on in a 1991 article, reprinted as a chapter in a 2006 book:

> Yet the exhaustion of the former epistemology resolves nothing because know-hows are no easier to study than knowledges. As soon as the latter are bunched in with the former, all problems reappear afresh, and our ignorance has not diminished . . . The profoundness of sciences comes from the fact that they definitively remove the possibility of direct, immediate, brutal access to the referent. That is even what makes them beautiful and civilizational. In an astonishing reflex, epistemologists give credit to exactly what the sciences are incapable of, that is, tearing us away from mediation to deliver us naked access, finally, to the referent. . . . we certainly conserve the specificity of scientific work – the guarantee of objective referents – without keeping anything of what the philosophy of sciences was supposed to give an account of.[32]

This is one of the rare occasions on which Latour makes positive use of a formulation from Gaston Bachelard, that of "workers of the truth," referring to scientists.[33] Latour is repeatedly surprised that many of those who call themselves materialists (Marxists in particular) are so reluctant to go looking in the field for *how sciences are made*, preferring to remain at the level of a general, merely definitional, discourse on objective truth.[34] All who do not rally to this point of view are relegated holus-bolus to the hell of "relativism." Latour keeps denouncing the *idealism of this materialism*, while praising a materialism that takes the question of matter seriously.[35] This is how he sets up *Irreductions*: "The materialism of this little précis should cause the pretty materialisms of the past to fade. With their layers of homogeneous matter and force, those past materialisms were so pure that they became almost immaterial."[36] Materials should not be taken seriously as homogeneous layers (like the social or the economic), but in *all their diversity*, which presupposes putting them to the test, specific tests, because "Whatever resists trials is real."[37]

According to many Marxists or critical sociologists, one certainly can carry out a sociology of scientific professions, but the specific relation of Science to Truth would make any sociology or anthropology of the sciences necessarily misleading, bringing about unfortunate, relativist – necessarily relativist – conclusions. Anyone wanting to go down this path would do nothing but join the army of the enemies of science, the army of "one thing is as good as another." They would even be part of the "rising tide of irrationality." The Great Divide must be upheld, encouraged, and all must be obliged to respect it, for fear of excommunication. Pierre Bourdieu defended this position tooth and nail, and announced his "antipathy" for "the phrasers and fabricators" [*les phraseurs et les faiseurs*] (i.e., Anglo-American STS folk . . . and Latour), while calling for a defense of the epistemological tradition of Bachelard, Koyré, and Canguilhem.[38]

Latour made fun of such critique, starting with the introduction to the French edition of *Laboratory Life*, his first book:

> The literature on the sciences is huge, but, as is the case with theology or apologetics in relation to religion, the assumption is made that science is taken for granted. Outside of this pious literature – a great

part of which looks like the *Manual for Inquisitors* – one can count on the fingers of one hand the few excellent books of the memories and analytical writings written by the scientists themselves ... As stimulating as these works might be, they are no remedy for the absence of inquiry, direct observation, and contradiction.[39]

The debate is political, not just analytical. Latour's opinion is that epistemology has a particular role to play in France (he even specifies that it relates to "French identity – in a very bad way, of course, at the moment"):

> It offers a uniquely French model for the resolution of political conflicts. There has to be a moment, when in Science (still with a capital S), whatever the material conditions are, or the influences, the intuitions, the passions, the false starts, objectivity and rationality emerge, and these are no longer linked with their conditions of production, no link at all: they are transcendent ... Nowhere is there such a profound link between the State and Science, not in Germany (despite Hegel), nor in Italy, England, America, or Holland: only France. If there is no radical way to escape the relative composition of influences, conditions, circumstances, relations of force, then there is no reason, no State ... Which is of course linked to the French version of universality, which is both scientific and political, what I would like to call national-rationalism.[40]

He comes back to this:

> Epistemologists are more interested in the State than in the sciences, especially in France where they want to establish the Republic! For them, science is but a way to talk about something quite different. Why do they leap at science like this? For the following reason ... we may have true statements – a few rare mathematical statements – that escape their conditions of production. Ever since Plato, we have been fascinated by these statements which allow us to circumvent politics.[41]

Stengers also takes up this proposition:

> Will Socrates be able to accept that it is his own questions about justice, the good, or love that have given rise to the discordant character of the answers? Will he be able to accept that those who have answered did not simply express "what they thought," but assumed a stance with regard to the situation that was proposed

to them? Will he be able to accept that it is the situation he created that produces the contradictions on which he bases himself, whereas Pericles, the politician he disqualifies, might perhaps have given rise to other responses, addressed to him in order that he might create their unlikely agreement?[42]

In order to demonstrate the extravagant character of Great Divide epistemology (endorsed as such, or, in many other forms, by a large swathe of the social sciences), Latour's method is very simple, to *expose to the eye* the way in which these researchers work, unpick in detail how a "fact" is never given, but has to be established. Already it is a matter of coming down to earth. So, the idea of the "fact" is the subject of a number of his commentaries, and he goes over it repeatedly in the course of his *oeuvre* in order to specify it better and better.

Two sites in particular are the focus of his fieldwork research, one in a contemporary North American research laboratory working on neuropeptides (TRF) and the other where he returns again and again to Pasteur's discovery. By going into the detail about how the sciences do what they do, Latour will thus work at "de-epistemologizing" the sciences.

This Great Divide will crop up again – as a principle in the "Modernist Constitution" (and thus a division of powers), he will say later – and in many other ways. It also concerns those who, in the past, emerged victorious (or vanquished) from a scientific controversy, such as that between Pasteur and Pouchet on spontaneous generation (this scientific debate was not yet settled; the biologist Félix Pouchet defended the theory of heterogeneity, but he ended up quashed by Pasteur's laboratory demonstrations).[43]

So Latour proposed investigating past scientific controversies as if we didn't know how they were resolved, who the victors were, and who the defeated. Now the other approach, adopted habitually by historians of science keen on epistemology, was to put on one side the victors who would be endowed with, as Descartes puts it, "a sound mind and a sound method,"[44] and on the other a whole list of factors, particularly "social factors," which would explain why those who lost out were in error. On the one hand, pure detached intelligence, on the other, social determinants obscuring a clear view: "[If] the social, psychological or economic determinations are only used to explain why a researcher was wrong, then they have no value."⁴⁵ He comes back to this point a number of times: "Error, beliefs, could be explained socially, but truth remained self-explanatory. It was certainly possible to analyze a belief in flying saucers, but not the knowledge of black holes; we could analyze the illusions of parapsychology, but not the knowledge of psychologists; we could analyze Spencer's errors, but not Darwin's certainties. The same social factors could not be applied equally to both."⁴⁶ "Most studies, in the history of science as well as the ethnosciences, bring in social, cultural, and circumstantial explanations only when the knowledge is deemed to be false. When it is true, it simply has no need for a social or cultural explanation; the truth is sufficient in itself."⁴⁷

The winners will be given the privilege of a sound method, while the runners-up wallow in an alienation of which they are obviously not aware. Latour strives to occupy a middle ground where he resorts neither to "social factors," nor to an overarching reason. Giving up the Great Divide means delving into the detail of sciences' results *in the making*, and also studying *already recognized* ones with the same tools and the same attention to detail.⁴⁸ *This is the restricted symmetry principle.* One cannot be content "to explain that victors in the history of science won because they were more rational or had better access to the nature of things."⁴⁹ Latour was explicit about this from 1984:

> All the failings of epistemology – its scorn of history, its rejection of empirical analysis, its pharisaic fear of impurity – are its only qualities, necessary if you want it to be a border guard. . . . All of which is to say that this précis, which prepares the way for the analysis of science and technology, is not epistemology, not at all.⁵⁰

Latour thus encapsulates the most urgent of tasks: "Describe, describe, and then describe some more."⁵¹

There are two enemies standing in his way, which he identifies in *Science in Action*. Not just epistemologists, but also relativists – which will not fail to surprise those who only have a rough idea of what his work is about (Latour, for his part, prefers to call himself a "relationist"). He accuses the former of short-circuiting, and erasing all the concrete work going into the establishment of "scientific facts," for the benefit of their results alone, their leading edge. He distances himself from the relativist point of view – which Stengers will call ironic – as it would consider results obtained by scientists as fictions among others that could only be maintained with the help of allies to the cause – colleagues, but also journalists, industrialists, etc. – through the influence of "the social," and therefore only taking human actors into account (researchers and their friends). He's been keeping the relativists at arm's length since 1984: "This is the weak point of the relativists. They talk only about forces that are incapable of allying themselves with others in order to convince and win. By repeating 'anything goes.'"[52]

But what does the success of researchers depend on if it is not the result of some connivance? This is where Latour give nonhumans their substance. Without his germs, Pasteur would collapse. It is only in association with them that he gains strength. Latour returns to the historical study done by Steven Shapin and Simon Schaffer, *Leviathan and the Air-Pump*,[53] in which Boyle, introducing a feather into a glass globe from which the air had been extracted (with the famous pump), demonstrated that Hobbes's ether did not exist. Boyle brought to light a new kind of witness: "inert bodies, incapable of will and bias but capable of showing, signing, writing, and scribbling on laboratory instruments before trustworthy witnesses. These nonhumans, lacking souls but endowed with meaning, are even more reliable than ordinary mortals."[54]

His encounter with Whitehead allows Latour to conceptualize what he has called the *principle of generalized symmetry*: "Whitehead's metaphysics allows us to take a decisive step in the philosophy of the history of science – blocked for some time on the question of the role that ought to be given to nonhumans."[55] It is not just the victors and the vanquished in the history of the sciences who should receive the same treatment (principle of restricted symmetry):

> linking humans and nonhumans, the principle of generalized symmetry . . . amounts to extending the notion of personhood to creatures of nature . . . The history of science becomes once and for all an existentialism extended to things. Nature, by becoming historical, becomes even more interesting, more realistic . . . Despite his hesitations, Pasteur does not dictate to the facts how they should speak. He mingles with them, offers a second chance to his

yeasts, sharing his history, his body, his laboratory, his cohort of colleagues.[56]

Latour adds, with humor:

> Not only does the yeast "happen to" Pasteur – transforming this honorable chemist into a world-renowned microbiologist – but Pasteur "happens to" the lactic yeast – transforming, through contact, this fermentation in the culture of the yeast nourishing itself on sugar. Yes, it has to be admitted, young Pasteur from Lille is a major episode in the destiny, in the essence, in the trajectory, of lactic yeast . . . By speaking of events defined in terms of their relations, I am sketching here the history of Pasteur and *his* yeast, of the yeast and *its* Pasteur.[57]

Latour brings to light, specifies, makes visible, the body of *actants* milling around the scientific scene; in the laboratories, the researchers are not *just* humans among other humans. When humans are successful in scientific laboratories, they are part of a cohort of other actors, not human, but oh, so very important, which they have found a way of calling forth. "Social constructivism," where things are decided only among humans, is nonetheless the path that was taken by a section of STS in the Anglo-American world. Stengers notes on this point:

> Latour has never stopped emphasizing that the scientist . . . cannot be "judged" in terms of social relations that would explain his success, because he is himself the active, enterprising co-creator of the "social" on which he depends. Latour thus asks that his sociologist colleagues learn to follow scientists, to describe how a scientist makes an ensemble of heterogeneous interests "hold together". . . .[58]

She underscores, as she is introducing an important theme in her work, the success of laboratory experiments without which everything else is up in the air: "Allies, resources, professional recognition, and the respect of the public are not enough. All of Joliot's work of connection would collapse if the neutrons did not respond to what he expects of them."[59] The researcher is like a mountain climber who has to find a good foothold in the course of a climb.

Latour enjoys drawing a general lesson for sociology from this, which becomes clearer after his dialogue with the ethologist

Shirley Strum: any sociology only interested in interactions *among human actors* could only be realistic for primate societies. "The sociology of simians, in this sense, becomes the limiting case of interactionism, since all the actors are copresent and engage in face-to-face actions whose dynamic depends continually on the reaction of others." It limits itself to copresent bodies. Simian societies are complex, those of humans complicated! Unlike humans, simians do not have an "operator of reduction," "intermediaries," "firebreaks." "Among humans, on the other hand, we actively globalize successive interactions through use of a set of instruments, tools, accounts, calculations, and compilers." Among humans, "objects are omnipresent in all the situations in which they are looking for meaning." "Reintroducing the objects, speaking again of the weight of things, according inanimate beings real social forces is for them [for sociologists] an error: the error of returning to objectivism, naturalism, or belief." And yet objects are well and truly actors. "Monkeys almost never engage with objects in their interactions. For humans, it is almost impossible to find an interaction that does not make some appeal to technics."[60]

Latour did not distance himself from STS all of a sudden. Later, he took himself to task for having misused the trope of the "social" in his first book: "I have used this trope in Bruno Latour and Steve Woolgar, *Laboratory Life: The Social Construction of Scientific Facts* [1979]. When this book was published, the failure of social explanation was not yet apparent. I drew its conclusions only later, by removing the word 'social' from the title of the second [Princeton, 1986] edition of the book."[61]

But how can we best represent these other "actants" that researchers mobilize? It is enough to look at how the researchers themselves go about it. Latour brings onto the stage "dissenters" that the researcher has invited to the laboratory, to show them "where the figure in the text comes from . . . The visitors have their faces turned towards the instrument and are watching the place where the thing is writing itself down (inscription in the form of collection of specimens, graphs, photographs, maps – you name it)."[62] The visitors are *told what to see*. "There is not much difference between people and things: they both need someone to talk for them. From the spokesperson's point of view, there is thus no distinction to be made between representing people and representing things."[63] Everything is happening as if the representative is saying "only what the things they represent would say if they could talk *directly*."[64] This is what Stengers is referring to when she speaks of "success": the experimental apparatus works "on a double register: it makes the phenomenon 'speak' in order to 'silence' the rivals." It allows the author "*to withdraw*, to let the motion *testify* in his place."[65] There are no free researchers who are at liberty to decide how to interpret what they are observing. What they have observed demands all of its space, and even that one speaks *well* of it.

But an observation made in just one laboratory isn't enough. Science is never carried out in isolation, even if it happens in the enclosure of the laboratories (where it is not done by a single researcher or by just one team), it has to be *taken up* by other colleagues in other laboratories. Latour shows how the solidity of a scientific "fact" is never immediate; a new "fact" often gives rise to disputes; a "fact" depends on the material it is drawn from and it is rendered "strong" or "robust" by all those who take what they need from it and adopt it for their own reasons, for their own benefit. But this is not a decision freed of all constraint. The "fact," or the new actant, must at each turn emerge victorious from every "test of strength," new experiments, new equipment against which it can demonstrate its robustness. The more successive tests there are, the more the "actant" impresses itself on a growing number of humans.

So the worst thing that can happen to a researcher's proposition is not so much being proved wrong as being ignored, to be neither taken up nor put to use, with no colleagues showing enough interest to put it to the test. Contrariwise, every new "test of strength" reinforces a "fact." All added up, these "tests of strength" sketch a "network," and a sociologist can map the collective that is being created in the same movement. This idea of network runs through all of Latour's work, as shown by the termite example that he evokes in one of his first books, which can be found again in one of his latest:

> For an entelechy [a word he is now using for actants] there are only *stronger* and *weaker* interactions with which to make a world. . . .[66] When one weakness enlists others, it forms a network so long as it is able to retain the privilege of defining their association. . . . Each network makes a whole world for itself, a world whose inside is nothing but the internal secretions of those who elaborate it. . . . we

could compare a network to a termites' nest – so long as we understood that there is no sun outside to darken its galleries by contrast. It will never be possible to see more clearly, it will never be possible to get further "outside" than a termite....[67]

This comparison will bear more fruit for Latour in 2021, when he uses it more generally to describe the thin layer where living things intermingle, what he calls the "critical zone":

> The thing that is indeed nice and practical about mushroom-cultivating termites and the way they live in symbiosis with specialized fungi able to digest wood – the famous *Termitomyces* which turns the digested wood into a nutritional compost that the termites then eat – is that they build vast nests of chewed earth, inside which they maintain a sort of air-conditioning system . . . The termite is in isolation, it's really a model of isolation, there's no need to say it; it never goes out! Except that *it* is the one that constructs the termite mound, salivating clump after clump.[68]

But let's go back to the enclosed space of the laboratory. There are never any "raw facts" waiting around for the researcher to pick them up. So all of the scientist's skill lies in their ability to invent a *stratagem*, a device that would allow the "fact" to appear and be stabilized over the course of successive tests. This is what designates the very start of the *network* that is going to get longer and wider with each test. The research will multiply the translation procedures that are going to assert there is a "fact." This might consist in a string of numbers, set out in tables, then repeated as graphs, just as we see in any number of scientific articles. These are *inscriptions* that are always different.[69] The steps are attached to each other by what they transport and have in common, something that circulates but remains similar, what Latour calls a *reference*, or an *immutable mobile*. How can we get a fix on this movement? "[T]he real phenomenon to concentrate upon is neither the ideas nor the images and what they might refer to, but is the trade-off between what is conserved and what is discarded when going from one scripto-visual trace to the next in line. This is . . . exactly the sort of necessity conservation transfer that deduction consists of."[70]

The chain of scientific inscriptions should always be *reversible*: one should always be able to work back to "the facts" that are undetectable or invisible under normal conditions . . .

whether it is hormones circulating in the brain (like the growth hormone studied by Guillemin at the Salk Institute), the lactic acid that Pasteur discovered, or even solar systems far off in space. Scientists are always interested in *distant* phenomena (so distant that a whole range of equipment is needed to make them present). One might need to work backwards along the chain of translations if, for example, other teams don't get the same results. Is there a point on the chain where there is a *failure* to translate? Since "data" are always "results," has the chain of successive outcomes been secured? Later, Latour summed up:

> This trajectory, made of discontinuous leaps, is what allows a researcher to determine that, for example, between a yeast culture, a photograph, a table of figures, a diagram, an equation, a caption, a title, a summary, a paragraph, and an article, something is *maintained* despite the successive transformations, something that allows him access to a remote phenomenon, as if someone had set up, between the author and the phenomenon, a sort of bridge that others can cross in turn. This bridge is what researchers call "supplying the proof of the existence of a phenomenon."[71]

It is a matter of "grasp[ing] continuity through a series of discontinuities,"[72] "an indefinite number of intermediate stations."[73]

This has a philosophical consequence which comes to confirm what I set out in the introduction to this book: "'Object' and 'subject' are not ingredients of the world, they are successive stations along the paths through which knowledge is rectified... It is the fact that 'occurs,' that emerges, and that, so to speak, offers you a (partially) new mind endowed with a (partially) new objectivity."[74] Latour can then write:

> If by "epistemology" we name the discipline that tries to understand how we manage to bridge the gap between representations and reality, the only conclusion to be drawn about it is that this discipline has no subject matter whatsoever because we never bridge such a gap – not, mind you, because we don't know anything objectively, but because *there is never such a gap*. The gap is an artifact due to the wrong positioning of the knowledge acquisition pathway. We imagine a bridge over an abyss, when the whole activity consists of a drift through a chain of experience where there are many successive event-like termini...[75]

There is no "dangerous leap," only a flux of experiences.

It is in this movement that the "fact" or the new "actant" emerging will multiply its attributes, sometimes changing its name (Pasteur's lactic acid became a gray mass of yeasty stuff, then it became a family of bacteria). All other researchers who work on new experiments are going to re-characterize this actant at the center of their preoccupations. They will add new qualities and attributes to it.

Relativists accuse Latour of giving too much reality to the actants that researchers bring to light, making independent partners out of them (when the researchers have done their work well), while they themselves fight tooth and nail to make them retain the status of "social constructions," refusing them any independence. While Latour shows how the sciences are peopling the world with new nonhuman actants, most of his STS colleagues want absolutely to maintain a world where only humans are acting and defining reality. This is what those accusing him of relativism have failed to see, even as he continues to insist: "In the world we inhabit, science is the socialization of beings invisible up to now, and who have very particular implications. We can domesticate them, master them, make them do things."[76] By "de-epistemologizing" scientific activity, Latour opens the door to its "ontologization" as a particular mode of existence (which will be the subject of chapter 7).

But we have to finish this chapter by going back to another conceptual advance of Latour's, without which one cannot understand the sciences as they occur: "black boxes." How can we account for the fact that the sciences are cumulative? Researchers do not need to always begin the same experiments over again. They can, at a given moment, after consolidating a whole pile of experiments using different methods, consider the existence of a "fact" or of a new actant as *brought about*. The debate is considered to be over when it is accepted by a sufficient number of scientific institutions (other laboratories, peer reviewers for journals or scientific conferences, academies, or learned societies), and that the fact or the actant can in turn become a *point of departure*. In future scientific articles, they will be cited merely by way of courtesy, in the introduction, with the classical formulation: "As X and Y have clearly established." It is not therefore an epistemological reason – knowing one's scientific mind is sound – that decides whether a black box should be closed, it is not because the researchers have respected preexisting proper scientific and objective criteria, that they have been able to convince their colleagues: it is simply the reproducibility of past experiments, their successful reprise, and therefore the lack of need to confirm and reconfirm, retracing one's steps and opening up again that which should henceforth have earned the name of black box. Objectivity is a result: "There is the question of objectivity, about which one must repeat what de Gaulle said about the economy: it will duly be taken care of."[77] Later, we shall see how Stengers made good use of this analysis in her attack on psychoanalysis.

Conclusion: "In other words, scientific activity raises no especially puzzling epistemological questions."[78]

Here Latour establishes a first distinction between science and technique, which have two different ways of establishing alliances "to stop a controversy,"[79] and for which the black boxes are not identical. Science always needs "fresh troops" who will mobilize once again what has become a well-established fact (for example, an identified and sequenced virus) and use it for their own purposes (as we have seen the worst that can happen is that no one, for their own reasons, makes use of it), find new attributes for it, and enrich it. Technique, for its part, has a need for "discipline in the ranks," for which the "indivisible whole" that has been created (a robot, a machine, an element of laboratory equipment, a vaccine) *holds*. This is engineers' work. Latour will later make this distinction bear fruit, essential as it is for ecological thought.[80]

With *techniques*, there are no *referents* or *mobile immutables*, but rather cumulative operations of enrichment. In an inquiry into the Aramis project for autonomous vehicles, which was the theme of a book written like a crime novel, Latour puts a sociologist into the story as the agent in charge of finding out who killed the project. He will find out in the end that there was no assassin.[81] The project died for want of love! The sociologist notes that a report

> presented the 1987 Aramis, word for word, as identical to Petit and Bardet [the originators of the] 1970 Aramis . . . Aramis had not incorporated any of the transformations that had occurred in

its environment. It had remained purely an object, a pure object. Remote from the social arena, remote from history; intact. . . . Look at the date: October 1987, one month before Aramis' death. Aramis has been exactly the same for seventeen years. The basic concept hasn't undergone any transformation, any negotiation . . . I'm beginning to understand. Yes, yes, there really is love in technologies . . . That's where the formal defect was, the sin, the crime . . . They didn't make Aramis a research project. They didn't love it . . . The whole thing should have been a research project. They abandoned technology while thinking that it was going to be finalized all by itself, that it was autonomous . . . that it had to be protected from its environment . . . They really succeeded in separating technology from the social arena! They really believe in the total difference between the two. To cap it off, they themselves, the engineers and the technologists, believe what philosophers of technology say about technology! And in addition, research for them is impossible, unthinkable; its very movement of negotiation, of uncertainty, scandalizes them.[82]

For seventeen years, nothing was remade, reconstructed, remodeled, as technologies demand. This difference between science and technology is made abundantly clear in the *Inquiry into Modes of Existence*: "Everything in the practice of artisans, engineers, technicians, and even weekend putterers brings to light the multiplicity of transformations, the heterogeneity of combinations, the proliferation of clever artifices, the delicate setups of fragile skills."[83] While Latour's ideas blur "definitively the notions of finalized research and fundamental research," the distinction between science and technology will be fruitful.

He even explains from now on that "The scorn with which people view technologies comes from the fact that they are treated according to the same model that we saw used to misunderstand the work of reference [which is to say the sciences]."[84] So while the experimental laboratory presupposes the purification of whatever enters it (in the way the laws of motion forget about friction), technology does the opposite:

Isabelle Stengers had the idea of undertaking a radical thought experiment to reduce all technological inventions to the "basic principles" recognized by scientists and presented to students as their "incontestable foundations": reduced to the Carnot cycle, locomotives would immediately stop running; limited to the physics of lift, airplanes would crash; brought back to the central dogma of biology, the entire biotech industry would stop culturing cells.[85]

This distinction between science and technology will, to Stengers' eyes, become essential, and she returns to it in 2002 when, in parallel with *Thinking with Whitehead*, she will ask questions of hypnosis and of therapeutic "techniques."

2
... Or Disamalgamate the Sciences

In 1989, at the very moment that Latour publishes the French edition of *Science in Action,* Stengers gives a lecture to psychoanalysts, the title of which must have at first seemed quite mysterious to them: "Scientific Black Boxes; Professional Black Boxes."[1] What led me to focus on this text, which might seem minor to some, is that here we find the author "situated" face-to-face with an a priori edgy audience. She talks about it in an interview:

> After I co-wrote *Order Out of Chaos* with Prigogine, I was a little taken aback by the interest psychoanalysts were taking, which had little to do with the type of interest in physics that we were hoping to raise. Among them, some started to see "bifurcations" and "dissipative structures" in the course of their analyses; a disquieting lack of taste that was evidence of their fascination with the "hard sciences." This was just transforming the difficult thinking of physicists into simple metaphors ... it was for the opposite reason I had gotten involved in the writing of *Order Out of Chaos*. I wanted people to stop paying homage to the authority of physics ... And here, all of a sudden, physics becomes more fascinating than before. Working with Chertok made me unpalatable for these people.

And she adds: "I was not very polite to the psychoanalysts."[2]
Why was Prigogine so important for her?

Prigogine was a famous physicist, but heretical because he wanted to rework a whole series of modes of thinking, at the heart of physics, that were part of the grand narrative of progress. In fact, Prigogine went against physics' own power, the right it gives itself to identify, behind appearances, a reality that does not distinguish the past from the future. As if one could not tell the difference between a film being screened forwards or in reverse. Prigogine dared to call that an absurdity.[3]

Absurdity? Make a note of this word; we shall come back to it.

The title of her lecture made explicit reference to the idea of black boxes advanced by Latour.[4] But where the latter had no need to differentiate (the difference between scientific black boxes and professional ones could be considered to be of no interest to him), Stengers puts her finger on the danger of generalization, specifically when it comes to psychoanalysis. So she is aligned with Léon Chertok, defender of hypnosis in the face of the claims made by psychoanalysis. If hypnosis interests her, it is primarily because she is trying to *impede psychoanalysis's desire to be scientific*.[5] This will be a real field of inquiry for her, both contemporary and historical.[6] "The desire to be scientific" is the main question of her text.[7] For a long time, Chertok had been battling against the idea of there being an "epistemological break," as it had been theorized by Louis Althusser, between hypnosis and psychoanalysis. Stengers will borrow from Latour but will insert a primary difference that will launch their entangled relation. This is how she puts it: "The fact that we are unable to define what a science is does not necessarily mean that this is a false question, one without interest. *Quite the contrary*."[8] "I have defined the scientific concept as always having two faces, one turned toward the phenomena which it is organizing the examination of, the other toward the scientists who are judged and ranked according to the types of interests they have invested in these phenomena."[9] What she advances is important, and she will later put it to use with the terms "requirements" and "obligations" (she also talks of the *bifacial* character of their concepts and their work): the requirements being "what the scientific collective demands of its environment to be able to carry on" and the obligations "the constraint one has to respect in order to belong to a collective of practitioners."[10] Thus one of the obligations experimenters have is to allow their competent colleagues to verify, reproduce, or deny their discoveries. It is in the terms of its obligations that a practice, scientific or not, can speak of its singularity. So she can say of Whitehead (who was a mathematician before becoming a speculative philosopher), "He thinks in the mode of obligation, as obliged by the problem he is constructing."[11] This is not a matter of searching for a demarcation criterion, as is the case for epistemology, but rather for a tool for inquiries addressing *practices*, a tool that will allow for a multiplication of approaches according to whichever science one is dealing with or, more generally, different knowledges. If practices are so interesting, it is because they are "always producing collective intelligence,"[12] therefore fragile, at risk of destruction. Writing originally in 1995, she makes this explicit: "Practice is first of all the manner in which we address ourselves to whatever it is we are dealing with, that is, the *requirement* that it satisfy certain criteria, and the *obligations* arising from the way it responds to this mode of address."[13] One could say that Latour, always simultaneously considering the two *actors* (or *actants* or agents [*agissants*]) at work in any scientific job – the researcher and the "fact" to which the researcher gives all its robustness – was not so far removed from this approach, but he did not turn it into a research tool, he hadn't "instrumentalized" it. Studying the work of scientists in their laboratories, he showed that the researchers make the actants emerge, as they in turn "make" the researchers "act" [*"font faire" les chercheurs*]: Pasteur's germs "make" Pasteur "design" the apparatus with which it is suitable for them to appear. This notion of *"font faire"* could well be Latour's version of requirements and obligations.

Stengers make this point repeatedly: "doing science," as in laboratories, is not something ordinary and generalizable, is not the result of the use of a method, however rigorous, but has the status of a rare *event*, an achievement that should be celebrated as such. "*Nothing has promised us anything*, and in particular, nothing has been promised to us such that, in any field of knowledge, the same type of event will re-occur." *Science is not a right of reason*, it is always a creation. Two things interest Stengers. First, not taking at face value everything that is presented in the name of science: "The event creates a distinction, which I think is crucial, between theoretical-experimental sciences, which, in every case, have 'made events' and pseudo-sciences, experimental psychology for example, which make the laboratory a place where scientific rationality claims the right to submit whatever it is interrogating to the status of experimental object."[14] One has to distinguish between the *modern* sciences and the *modernist* ones, the sciences "which are passionate" and those which are "sad and stultifying."[15] She takes the example of the neurosciences: "A good number of neuroscientific demonstrations contribute only to an accumulation of the kinds of 'facts' that are of no use to working colleagues – even if they are candy for the media."[16] She suggests greeting this kind of work "with mocking laughter, or an amused indifference."[17]

As Latour has shown, a statement emerging from an experimental laboratory should be reproducible by other researchers in other laboratories, which allows it to stabilize. But this "[r]eproducibility in the experimental sense doesn't have much to do with what, by contrast, is called repeatability: every time the head of a human or other vertebrate is chopped off, they die."[18] Stengers will continue to maintain this distinction, between sciences and pseudo-sciences, in all her *oeuvre*, but it does not overlap at all with the rationalists' distinction – the followers of Method. On the contrary, she will use it to rehabilitate knowledges that are habitually eliminated by rationalists, and show the inconsequentiality of a good number of practices that are glorified as "scientific," with the full support of those for whom only "Method" can count. In this way, no one can profit from her "critique of psychoanalysis" to bolster behaviorism, for example, or other so-called scientific practices, or, on the other hand, find a better justification for rejecting whatever is considered to be charlatanism. Her tour de force is to never allow this type of generalization, and anyone who tries this kind of maneuver risks regretting it very quickly. Latour was able to say, in *Irreductions*, that "the sciences" were "[S]parse and fragile, and above all sparse."[19] And he is even closer to Stengers when, in 2012, he declares, "This is the project of the exact sciences: they are exact, not because they are mathematical, but because they invent equipment that has this *unique adequacy*, as Garfinkel would say. The *unique adequacy* means we have to find a piece of equipment that will work for that particular object, but not for another."[20]

So "scientific black boxes" and "professional black boxes" shouldn't be confused, even when they try to adorn themselves with the same finery, the latter trying to trade off the advantages and the prestige of the former. Because that always happens at the expense of other practices. Such as in the case of psychoanalysis, those that are discounted under the name of practices of *suggestion*. So what is a professional black box?

> Why would analysts want to put a question mark over the historical privilege that they find in their heritage? Contrary to scientists, for whom their professional status depends on the networks they participate in or help build . . . analysts depend only on their clients. They are what can technically be called "small independent" businesses. In this perspective, the idea that analysis is something special is defended, rather than demonstrated, since it is from this that the influx of clients comes, at least in part.[21]

If Stengers denies psychoanalysis's pretensions to be a science, and refutes the idea that *rationalist* procedures would allow a field of knowledge to gain this status, she can, in the same way, give room to ways of proceeding in fields of knowledge where the event that characterizes the sciences often called hard has never been *produced*, if these fields know how not to imitate or repeat, but to escape domination and invent new ways of distributing "requirements and obligations." On this, she cites experts on evolution who "succeed in learning, discussing, the partial, local reliability of observations that are not *proofs but clues*."[22] Linking back to Ginzburg. Latour will not rally to this point of view, even though he will arrive, albeit in his own way, on a different path, at a distinction between fieldwork science and laboratory science, and not give the name of "science" to either psychology or, later on, economics.

In conclusion, Stengers stresses the consequences of her idea for psychoanalysts:

> The narcissistic wound described by Freud does not have any impact on shamans, mystics, or thaumaturgists, only on westerners, white ones descended from Descartes. Is it not a paradox that although the Freudian technique did not live up to the hopes of its creator, psychoanalysis confers on white people the extra power needed not just to know nothing about the practices of those folk, but also to cast theoretical judgments over them?[23]

So she is not condemning psychoanalysis for not having succeeded in turning itself into an experimental science. It is its unfortunate pretensions to *do science* [*faire science*] that in fact hinder it, that have prevented it from realizing its full inventive potential. Stengers is Latourian on this point. Casting oneself in the wrong light is not just an unfortunate detail to be corrected but a catastrophe from the point of view of what she will later call "ecologies of practices," the making of a formidable blindly destructive machine. Her position does not in any way open out onto praise for other modernist psychological techniques: these often share, albeit in different ways, the illusions (or naiveties) of psychoanalysis. Later, with Tobie Nathan, she will extend these reflections, and we shall return to them.

So, while both our authors are interested in scientists' *practices*, they go about it differently. Latour takes up the methods of ethnographic inquiry (more than anything sociological) and rejects any distinctions that would risk bringing epistemology in through the window after he had thrown it out of the door: "They are skeptical and unbelieving about witches and priests but, when it comes to science, they are credulous. They say without the slightest hesitation that its efficacy derives from its 'method,' 'logic,' 'rigor,' or 'objectivity.'" Believing in epistemology puts the cart before the horse:

> Those who call themselves "scientists" always put the cart before the horse when they talk, though in practice they get things the right way around. They claim that laboratories, libraries, meetings, field notes, instruments, and texts are only *ways* and *means* of bringing the truth to light. But they never stop building laboratories, libraries, and instruments in order to create a focal point for the potency of truth . . . A purely scientific science would rid us of scientists. For this reason they are careful not to kill the goose that lays the golden eggs.[24]

For her part, Stengers also deals with practices, but from within philosophy, having benefited from the study she did with Chertok on the classical case of psychoanalysis. This is not a matter of looking for epistemological criteria but of taking into account the nature of *requirements* and *obligations* of each practice. This is something she stresses throughout her *oeuvre*: if one turns what happens in the experimental laboratory inaugurated by Galileo into a general law, then one is "shifting the burden of proof from the register of the event to the general register of judgment. . . . Science here is defined as being *against* opinion, the burden of proof is weaponized by a general tribunal that, above all else, has the mission of establishing a pedagogic relationship with a public henceforth defined as credulous."[25] Exit, epistemology!

> Scientists, they say, have objectivity as their common "worth", and this might actually be the sole claim that could bring together practices as diverse as physics, sociology, psychology, or history . . . This informs, by the way, the content that Gaston Bachelard attributes to scientific rationality: an ascetic "no" directed toward a veritable gallery of horrible opinions. As Bachelard puts it, "By right, opinion is always wrong, even in cases when, in fact, it is right."[26]

Stengers draws an example from her field of choice: for her, chemistry, in the nineteenth century, did not say "no," as Bachelard would have liked, to the chemical arts. The "modern" academic chemist no longer had anything to say to artisans because she had no way of understanding them. Her knowledge now related to procedures that presuppose standardized reagents that the industry has put on the market (a chemist on a desert island would be stripped of any relevant knowledge). There was no "epistemological break," rather a disconnection, as "the only true interlocutors for the new academic chemists, the only ones who understood their language, were now those who inhabited the industrial world also in the making."[27] We shall come back to this.

3
A Brief Exercise in Empirical Philosophy

I thought it would be interesting to illustrate the difference, or the complementarity, of Latour and Stengers' approaches through an example the former developed at length in an article published in 1993 after he went on a research trip to Boa Vista, in Brazil, with a team of researchers. What they were trying to determine, at a precise location, was whether the Amazonian forest was advancing, or retreating and giving way to the savanna.[1] The team comprised researchers from different disciplines: a botanist, two pedologists (soil scientists), and a geographer (geomorphologist).

What Latour was aiming to do with this article that he considered to be an exercise in "empirical philosophy" is the decomposition of the different stages of the work of the team in order to bring to light the precise chain of operations that the material studied (the soil and the vegetation) will undergo as it transforms through successive translations.

> Between the sandy savanna and the clayey forest, it seems that the twenty-meter-wide strip of land spreads out at the border, on the savanna side. This strip of land is ambiguous, more clayey than the savanna but less so than the forest. It would appear that the forest casts its own soil before it to create conditions favorable to its expansion. Unless, on the contrary, the savanna is degrading the woodland humus as it prepares to invade the forest.[2]

How is it that the world of *things*, spreading out in a more or less formless way before the eyes of the naive observer, can become a *sign*? For Latour, this is a matter of "understand[ing] more concretely the practical task of abstraction."[3] At the start, the pedologists remove lumps of earth from different places that have been well mapped out in the area to be studied, and deposit them in a series of cardboard cubes. They are all coded so that they can later identify what sample they are dealing with:

> Consider this lump of earth. Grasped in René's [one of the pedologists] right hand, it retains all the materiality of soil . . .Yet as it is placed inside the cardboard cube in René's left hand, the earth becomes a sign, takes on a geometrical form, becomes the carrier of a numbered code, and will soon be defined by a color [according to the Munsell code]. In the philosophy of science, which studies only the resulting abstraction, the left hand does not know what the right hand is doing! In science studies, we are ambidextrous: we focus the reader's attention on this hybrid, this moment of substitution, the very instant when the future sign is abstracted from the soil. We should never take our eyes off the material weight of this action.[4]

Each sample will also be studied and classified according to its texture, its particle size. "A word replaces a thing while conserving a trait that defines it."[5] Latour uses photographs to allow his reader to follow all these intermediary steps, these successive translation procedures, in order to make them as concrete as possible.

Through all these transformations that succeed each other in an orderly fashion after the sampling, one should be able to grasp "the constant that is maintained," what he calls *reference*, as we saw in the last chapter: "It seems that reference is not simply the act of pointing or a way of keeping, on the outside, some material guarantee for the truth of a statement; rather it is our way of keeping something *constant* through a series of transformations."[6] Each stage is retained, including the first (the pedologist returns to Paris, his bags full of samples, the lumps of earth), in order to verify that no mistakes were made in the course of the procedures. And they have to be *reversible*, therefore verifiable (and not, in this case, reproducible) by other teams: "In none of the stages is it ever a question of copying the preceding stage. Rather it is a matter of *aligning* each stage with the ones that precede and follow it, so that, beginning with the last stage, one will be able to *return* to the first."[7] One thus never notices any break between things and signs:

> Notice that, at every stage, each element belongs to matter by its origin and to form by its destination; it is abstracted from a too-concrete domain before it becomes, at the next stage, too concrete again. We never detect the rupture between things and signs, and we never face the imposition of arbitrary and discrete signs on shapeless and continuous matter. We only see an unbroken series of well-nested elements, each of which plays a role of sign for the previous one and of things for the succeeding one.[8]

In this way we "oversee and control a situation in which we are submerged, we become superior to that which is greater than us, and we are able to gather together synoptically all the actions that occurred over many days and that we have since forgotten."[9] Is the diagram we end up with "a construction, a discovery, an invention, or a convention? All four, as always."[10]

And Latour concludes:

> We have taken science for a realist painting, imagining that it made an exact copy of the world. The sciences do something else entirely. . . . Through successive stages they link us to an aligned, transformed, constructed world. We forfeit resemblance, in this model, but here is compensation: by pointing with our index fingers to features of an entry printed in an atlas, we can, through a series of uniformly discontinuous transformations, link ourselves to Boa Vista . . . I can never verify the resemblance between my mind and the world, but I can, if I pay the price, *extend* the chain of transformations wherever verified reference circulates through constant substitutions. Is this "deambulatory" philosophy of science not more realist, and certainly more *realistic*, than the old settlement?[11]

Here, Latour has been able to demonstrate the same thing he did ethnographically in Guillemin's laboratory, where they were trying to identify the growth hormone.[12] In order to do this, he has had to assume that what he was dealing with at Boa Vista was a "laboratory," but in the open air: "For the world to become knowable, it must become a laboratory."[13] Laboratories "are excellent sites in which to understand the production of certainty, and that is why I enjoy studying them so much."[14] So the laboratory is not always the "experimental laboratory" of Pasteur, or of Guillemin; the idea can be extended to fieldwork sciences, and the separation between the two can only obscure what it is he wants to show.

In the end, Latour has wanted to stress two things that are good at distinguishing sciences-as-they-are-carried-out from epistemological pretensions: if all the stages of scientific work are reconstituted carefully, then *construction* and *fact* are no longer in opposition. All of the translation procedures carried out in the laboratory (taken broadly), cause an *immutable mobile* to circulate designating a "chain of reference."

Even if it was not his intention, Latour's article brings to the fore the *singularity* of the question to which the team of researchers seeks to respond, without having a model to follow which would assure the correct answer.

Stengers could weigh in here in several ways: the importance for this kind of research of the clues that Ginzburg spoke of, which appears at the beginning of the article in the remarks Latour made about his first photo:

> Little figures lost in the landscape, pushed off to the side as in a painting by Poussin, point at interesting phenomena with their fingers and pens. The first character, pointing at some trees and plants, is Edileusa Setta-Silva . . . [she] is pointing to a species of fire-resistant trees that usually grow only in the savanna and that are surrounded by many small seedlings. Yet she has also found trees of this same species along the edge of the forest, where they are more vigorous, but do not shade any smaller plants . . . Edileusa hesitates . . . [she] believes the forest is advancing, but she cannot be certain because the botanical evidence is confused: the same tree may be playing either of two contradictory roles, scout or rear guard.[15]

Hesitations are a necessary part of the researchers' work.

The question the researchers are trying to answer is at the heart of an *interdisciplinary procedure*, which does not aim to show how well the researchers get on with each other but rather it is indispensable for answering the question which brings them together. There is no other way they can work. The botanist needs the pedologist, and they both need the geomorphologist. And the conclusion of the inquiry will be that the hesitation must be maintained; one is going to have to appeal to other unknown specialists, those who know about "earthworms, whose activity on the studied site we have been able to verify . . . The study of this worm population and the measure of its activity will therefore supply essential data for the continuation of this research."[16] This requirement for interdisciplinarity and for recruiting researchers from other disciplines (biologists) is part and parcel of what the question at hand *demands*. All the researchers from all the disciplines have to learn how to work, and to hesitate, together. This is yet another illustration of what Latour calls the *principle of generalized symmetry*: the earth, and the research team, each have a history. There is no "subject" observing an "object," but two entangled "beings." "We can put it even more simply: can philosophy be forced at long last to count beyond one or two (subject and object) or even three (subject, object, and going beyond subject and object through some dialectical sleight of hand)?"[17] Once again, we come to the difficult issue that I have been wanting to discuss ever since the introduction to this book.

This is not a matter of the work carried out by the research team being overridden by another team. Here we are not waiting for some *event* that would turn the way of conceiving the problem on its head. In any case, the team has not been able to conclude but this does not mean it has failed. It is not a matter, as in experimental laboratories, of "shutting up our rivals." Rather, other work will come along that *adds* to theirs. Latour's formulation of this is that the *network* that is spun around these inquiries is not made up of contrarians trying to show that the first team has merely fabricated an artifact; rather, it is a network of collaborators spreading ever further and becoming more inventive. The reply to the question – is it the forest or the savanna that is advancing? – does not establish a generalized scientific "fact," its ultimate success is not to shut a black box; it is only valid in that particular time and place. And a bit further afield in the Amazonian forest, the answer might be a different one. The researchers "learn how to describe it from what they observe."[18]

It could very well be the case, even if Latour has not envisaged it in his study, that the researchers have a need to escape from what he has defined as a "laboratory" in order to appeal to the knowledges of those who inhabit the region: what do they know about the history of this intermediary zone between the savanna and the forest? What could their parents or grandparents have told them about its past? It could well be that (here or elsewhere) inhabitants have an answer to the question as to whether it is the forest or the savanna that is advancing. Conversely, it could be that the knowledge gathered by these researchers is of interest to these same inhabitants for their own reasons: their relations with the big landholders, for example, or with government. It might also be of interest to ecological activists, and all those worried about the future of the Amazonian rainforests.

One might well ask oneself about what happens to this kind of research in the context of the new knowledge economy that "designates a strong reorientation of public research policy, making partnerships with industry a crucial condition for the financing of research."[19]

This team is not imitative, it is inventive. It proves its imaginative and creative credentials by trying to answer the quite specific question that is asked of it. Reproducibility is not at stake, as in the experimental laboratory, nor are the researchers repetitive like those who think it is enough to get the methodology right. That, too, has to be invented and permanently modified in the course of the research work (for instance, when it is realized that the earthworm specialists will have to be called upon . . .). It is a case of learning afresh. What this research team learnt at Boa Vista could nevertheless be of use for other teams working in other parts of Brazil or on other tropical forests under threat. Science, here, is certainly an adventure rather than all-conquering reason. As Stengers has said:

> I have tried to characterize scientific practitioners (in contrast to those who serve Science) as gathered together by a "common," that is to say, by a cause: they are *engaged by a type of achievement* proper to each field, *the eventuality of which places obligations on those who belong to this field*, forces them to think, to act, to invent, to object, that is to say, to work together, depending on one another.[20]

Latour had already engaged with a field science when he met Shirley Strum, the baboon ethologist. He taught with her for a few years, beginning in 1979. Later on, he wrote:

> I knew laboratories, of course; I was beginning to measure the artificial aspects of experiments . . . but how was I to characterize the space created by baboon troops that were being followed by researchers? *Followed* and not *preceded* by them, this says it all. . . . And yet these researchers capable of following and not dominating

their object of study were producing science, and very good science at that . . . I began to imagine other relations between the course of knowledge and that of the known world.[21]

This was a contrast already brought out by Deleuze and Guattari in *A Thousand Plateaus*:

> Royal science, with its axiomatic and theorematic power . . . put to the test by nomad or ambulatory science, which invents problems, which follows "connections between singularities of matter and traits of expression." Nomad or ambulatory sciences have no age or geographical attachments. The solutions to the problems they invent are, moreover, in a reciprocal presuppositional relationship with the collective activities of those who are always living in one territory or another.[22]

Both Latour and Stengers call upon scientists to learn to present themselves well. Instead of appealing to an epistemology, or to the rule of reason, instead of denouncing the growing threat of irrationality, these scientists could mount more fruitful objections to relativist descriptions. Stengers gives their protest the quality of fabulation:

> the manner in which you describe us, as addressing a reality that is incapable of differentiating between us, is destroying us! We are no longer anything without an association with this nature that you are characterizing as mute, as radically incapable of playing the role that we appeal to it to play, that is to say, as incapable of making the possibility of our agreement anything other than an agreement among humans![23]

If being a sociologist of science were a matter of explaining the sciences by way of their "social context," then Latour is obviously not one of them. He could even be their most formidable enemy, which is something they themselves have often considered. It is a pity that many of his other critics, especially in France, have ignored it. He progressively distanced himself from the sociology of science, but this led him to do the same with sociology itself, especially so-called "critical" sociology, and he came to define his research work increasingly as more related to ethnography, which was better equipped to take on the Great Divide. If neither society, nor the social, could explain the work of scientists, their way of closing their black boxes, what then was the good of these notions of society or the social? If there are no "social constructions," what is the social, what is society? A new field of inquiry opens up. Perhaps it is not just the sciences that have been badly described, but society and nature as well! The stakes are considerably higher, and this became the object of an entire book, first published in English in 2005: *Reassembling the Social*.[24]

4
Sociology or Politics?

Quite early on, Latour wrote, "If *sociology* were (as its name suggests) the science of *associations* rather than the science of the social to which it was reduced in the nineteenth century, then perhaps we would be happy to call ourselves 'sociologists.'"[1] With this statement, he is distancing himself from Émile Durkheim's sociology in favor of Gabriel Tarde's, who has the advantage of:

> dissolving all structures – that of the pure and perfect market, of course, but also those of the social world which are accepted by sociologists ... Along with the dissolution of society, all the metaphors of embeddedness also disappear. Economics no longer lies in the Procrustean "bed" of the social ... because there is no more bed, no more pillow to rest one's head, no more canopy, no more duvet.[2]

The "facts" established in the scientific laboratory (after a long successful labor of "fabrication"), or the technical objects, can only rapidly wither away should the networks collapse (like the mobile phone becoming useless when the network goes down), networks to which they gave birth and in which they are inscribed at the very moment they create them. These networks are extended via new laboratories that look after, and cultivate, the "facts" in all their relevance.[3]

But networks are not just a string of laboratories; they are comprised of many other institutions: private enterprises, diverse agencies, etc. "If science grows, this is because it manages to convince dozens of actants of doubtful breeding to lend it their strength: rats, bacteria, industrialists, myths, gas, worms, special steels, passions, handbooks, workshops, etc. A whole *cour de miracles*, a crowd of beggars performing their handicaps."[4] Networks pay no attention to the boundaries separating different disciplines. Is not any boss of any laboratory constantly on the hop from one field to another? Heading off to the lawyers to consult about a possible patent? Or to a publisher to popularize the research and gain respectability? To researchers in another field to see if what they have discovered could be useful in a totally different perspective?

In his ethnographic inquiry, Latour has covered in detail the way in which the boss leaves the laboratory to explore the world looking for quite disparate allies. "Tracking the movements of a researcher could be an exhausting thing to do, leading the ethnographer to *visit* all sorts of countries around the world and to hang out with many more groups of people in society than they might expect to: highly placed officials, CEOs, university staff, journalists, religious figures, colleagues, etc."[5] Scientific networks are permanently extended; this is even the capacity that distinguishes them from all others. As Latour adds, "to extend the field of validity of their results [the method adopted by scientists] consists in the construction of standardized networks. Now, the extension of a standardized network of proofs does not, in principle, give rise to any difficulty: universality is accessible, but at the cost of extending the network."[6]

Latour will later add two other elements to that of the network: accumulation and centers of calculation. Scientific statements that are circulating in ever-extending networks are collected in "centers of calculation," which allows for them to be reorganized, aggregated, and made useful for other researchers. Immutable mobiles thus become *combinable*. But is it possible to generalize this analysis deriving from the experimental laboratory for all practices presented in the guise of science? Latour expresses a doubt when he writes in a footnote which we have already cited: "One can read with interest Ginzburg's (1980) counterexample which, he thinks, can separate the *sciences* of the trace or symptom from 'exact' sciences."[7] He then elaborates:

Sociology or Politics?

> This does not mean that "theories" simply follow the accumulation of "data" – on the contrary, "mere stamp collecting" is often opposed to "real science" – but simply that any a priori epistemological distinction between the two makes the study impossible. The problem is that we lack independent studies on the construction of this contrast between "data" and "theories." For such an endeavor on the relations between physics and chemistry, see I. Stengers (1983).[8]

But he offers no further elaboration, as if the occasion for it never arose . . .

Since a "fact," as inscribed in a long network, is in itself evidence that it has resisted multiple trials of strength, we may be tempted to say that we are dealing with a scientific fact: "My contention is that on the contrary we must eventually come to call scientific the rhetoric able to mobilize on one spot more resources than older ones."[9] Of course, he had earlier written: "No matter how impressive the allies of a scientific text are, this is not enough to convince. Something else is needed."[10] "New objects become things, existences become essences, laboratory work that construct them starting with a list of trials becomes a discovery of *what* was always already there . . . as soon as they move away from their laboratory conditions of production, they become named things that appear independently of the tests via which they have demonstrated their strength."[11] Existence precedes essence. This is also a red line to be found in the *Inquiry into the Modes of Existence*. And Latour has often come back to it in order to improve the formulation. The first version, provisionally, is in *Laboratory Life*:

> Scientific activity is not "about nature," it is a fierce fight to *construct* reality. The *laboratory* is the workplace, and the set of productive forces, which makes construction possible. Every time a statement stabilizes, it is reintroduced into the laboratory (in the guise of a machine, inscription device, skill, routine, prejudice, deduction, program, and so on), and it is used to increase the difference between statements. The cost of challenging the reified statement is impossibly high. Reality is secreted.[12]

And, in a footnote, he elaborates: "If reality means anything, it is that which 'resists' (from the Latin *res* – thing) the pressure of a force."[13]

When Stengers wrote *The Invention of Modern Science* (in French) in 1993, she had immediately grasped the risks run by sociologists of science, with which Latour is often identified; she saw the "science war" coming, three years before the Alain Sokal hoax, when she opened her book with this premonition: "A disturbing rumor has been spreading in the world of scientists. It seems that there are some researchers – specialists in the human sciences, no less – who are challenging the ideal of a pure science."[14]

She is therefore slowing down Latour's move by taking on new methods, what she will call a "politics of science." Yet she is not in any sense an ethnologist or sociologist, rather she presents herself as the inheritor of the philosophy of science, with a predilection for Popper (but it is not his so-called "falsification" theory that interests her[15]); nor is she an "epistemologist" because she always emphasizes that it is practices she wants to account for and characterize. If Bachelard and Canguilhem take up little space in the first sixty pages of her book, it is rather that she is doing careful work on what Popper, Kuhn, Feyerabend, and Lakatos have to say.

She starts with this distinction: "In order to make scientists actors like any others in the life of the city (the 'political' preoccupation), it is not necessary to describe their practice as 'similar' to all other practices (the 'sociological' preoccupation)."[16] The "detailed" description of sciences in the making (or already made) should not turn them into *banal* practices. Where a sociologist might be tempted to erase the specificity of the work of scientists to turn them into actors in the city like anyone else, politics can take another route. In order to do that, she has to turn to what we may call the privileges afforded by the fact of presenting oneself as a scientist. The definition of "science" is never "neutral" because "from the beginning of so-called modern science, the title of science has conferred certain rights and duties on those who call themselves 'scientists.' Every definition excludes and includes, justifies or puts into question, creates or prohibits a model."[17] So, even putting aside the examples chosen by Bachelard to illustrate his theory of "epistemological obstacles," what should be interrogated is his "extraordinary influence in French epistemology . . . the strategic function it played in domains he himself did not tackle,"[18] as we saw for the field of psychoanalysis.

Stengers takes Galileo's inclined plane as her starting point. The device Galileo created "does not bring the movement of falling bodies into 'existence,'" as Robert Boyle did for the void, but "designates it in its novel singularity."[19] Thanks to the inclined plane, this movement becomes able to establish *the way in which it should be described* as a matter of fact, and thus to exclude the philosophical idle chatter about "reasons." Stengers will later add humorously, like Whitehead, "this conflict places modernity under the banner not of a progressive history, but of a revolt: a 'historical revolt' against medieval rationalism."[20]

This foundational example will allow Stengers to promote the idea of "theoretico-experimental" sciences: "theoretical production is expected and legitimate within the practice of the modern sciences."[21] In these types of cases, the phenomenon staged *"has testified to its truth."*[22] It has silenced all the other rival "fictions" that might also claim to explain it. The link between science and rationality henceforth heralds "a new use of reason." All of a sudden, passing through Stengers' hands, the "laboratory" that links experiment and theory takes on a whole new singularity, as long as this link is placed under the sign of the event, a success – not a method. And this type of event taking place in a laboratory is something rare that therefore deserves to be celebrated. As she will say quite strikingly in a later book, *La Vierge et le Neutrino*, when something like this happens, "scientists dance in their laboratories."[23] It is a moment when experimenters like to say that "nature has spoken."[24]

However, she later makes the point that if Galileo is the founding father of experimental laboratory science, this success brought with it an unfortunate turn of phrase, a kind of early version of epistemology: "he was the first to promote the general, unilateral authority of science, conquering the world, defining what really matters and what are mere illusory beliefs, thus giving his blessing to the destruction of innumerable other ways of relating, knowing, feeling, and interpreting."[25] Later, she will even call this *"one of the most successful propaganda operations in human history*, as it has been repeated and ratified, even by the philosophers whom Galileo stripped of their claim to authority."[26] And the operation was all the more remarkable in that physics, emerging from Galileo's laws of motion, was not just the first experimental science in the chronological sense, it is also the one that will be recognized as the only one competent in the "discovery" of the laws of the universe. Thus, Stengers writes, as we recall that she started off as a chemist, "Physicists like to repeat, with respect to chemistry, that chemistry learns, while physics understands. Indeed, chemists 'learn from' the dizzying variety of chemical compounds how to characterize what their ingredients are capable of. Intelligibility, when it occurs, is obtained after the facts: chemists *render intelligible*."[27]

Stengers comes back to this a number of times: "I do not aim to diminish the grand success of the Galilean-Newtonian paradigm. I prefer to leave such success to its beautiful solitude and fragility. Its realization of the coincidence between intelligibility and submission is relative to a rarified milieu, purged to the utmost of what can never be entirely eliminated, *friction*."[28]

Stengers will underscore a second point, which is part of the characterization of the sciences now being designated "theoretico-experimental": to gain entry to a laboratory, a phenomenon should be *extracted* and *purified*. This is what Galileo did in eliminating the question of friction in order to write his equation on falling bodies. That is what Pasteur did to bring germs into existence.

Stengers, in this way, gives herself the means to sustain what will be a common thread running through her *oeuvre*, meeting up with Latour's refusal of rationalism: all of the sciences are not scripted to the same model. One has to avoid like the plague any "submission to a method which is supposed to guarantee objectivity or the validity of its statements."[29]

As Stengers predicted, many of the scientists who had welcomed Kuhn's "paradigm" idea universally rejected the proposals of sociologists of sciences, making no difference between the relativists and Latour: they read such proposals as denying the specificity of their practices and what they considered their special relationship with the truth. They felt insulted. Stengers' discussion of Latour's ideas is located precisely there – where the scientists thought they saw a "declaration of war" – with the aim of solving this problem while respecting what she called the Leibnizian imperative: do not shock common sense. Which means here: take seriously the "scandal" experienced by the scientists.[30] So Stengers had to fight on two fronts: against a certain kind of sociology of the sciences, one that could be called relativist, and equally against rationalist positivism.

I could try to sum up by saying that while sociology tries to show, to bring to light, to render *visible* (Latour says it in a number of ways), "Stengersian" politics immerses us in controversies, debates: it lets the actors speak. It listens to them. So it is not possible to ignore the scientists' protestations. This breakdown of relations with the scientists who were scandalized by the way they were being described matters. Furthermore, it urgently needs to be repaired if possible. Because there is a danger lurking, the worst being that the scientific community ups sticks and shifts allegiance definitively to the side of a rationalist vision of the sciences. This would allow all those hiding behind science to bring off a victory for their political projects, rationalist projects in particular, the principal aim of which is to eradicate all "nonmodern" knowledge practices. This can only reinforce the Great Divide. Nothing could stop open war. The issue therefore goes beyond epistemology alone. If it were only philosophical theories under threat, the danger would not be so great. But not listening to the cries of protest coming from scientists would mean depriving the city of their considerable contribution in favor of those who speak in their name without any qualifications to do so. Scientist *should be allies*, but for Stengers they have to learn to present themselves differently. So it is catastrophic to throw them into the arms of the rationalists and epistemologists when things could have been done differently, as Stengers learnt in Prigogine's laboratory. This is why Stengers offers a "politics of knowledge," or a "political invention of the sciences." The aim is one in which "the words 'opinion' and 'reason' lose their power of self-definition by being opposed to each other."[31]

> The same question presents itself with regard to the person who claims to speak for others as it does with regard to the theory that claims to represent the facts [including the sociologist's]: "How does one recognize the legitimate claimant?" We can, in this sense, speak of the birth of a politics of knowledge and of a science of politics at one and the same time.[32]

We have to begin with the distinction between "who have the right to intervene in scientific debates (the right to propose criteria, priorities, and questions) and those who do not have this right."[33] This "political" dimension is constitutive of the sciences. And it is right there that the opposition "of scientists to any sociology of the sciences can then be understood in *political* terms."[34]

How, then, is the contribution of scientists best represented? How can one propose a description of their work which recognizes their specificity without reference to an overarching rationality or a "method" for all occasions? In other words, without agreeing with the rationalists (or the epistemologists)? Stengers has a formulation that designates the success of the experimenter (not to be confused with all those posing as "scientists") and records its achievement. How can we characterize the events that might happen in an experimental laboratory? Stengers uses one of Latour's ideas on the relation that researchers have with the beings that they bring into being in their laboratories. He suggested considering them as spokespersons (mute spokespersons!). Let us remember what he said: "the spokesperson is seen not really as an individual but as the mouthpiece of many other mute phenomena"[35] "or, even better, the spokespersons are considered mutes via which the represented entities speak without hindrance."[36]

Stengers gives it a precise definition – which could be called political – in terms of *power*: "This is the very meaning of the event that constitutes the experimental invention: *the invention of the power to confer on things the power of conferring on the experimenter the power to speak in their name*."[37] Which is actually the case with Galileo: "If there is one case which may claim being a vector of novelty, it is indeed the case of the twofold 'production' of a Galileo discovering that he is capable of telling the difference between fiction and scientific statement, and a new type of data ('experimental') that confers this capacity upon him."[38] This is a landmark proposal, and Latour will take it on board, particularly for speaking about Pasteur, without however going as far as accepting Stengers' project to use the specifity of the experimental success in order to disamalgamate the sciences: "Scientists speak in the name of phenomena to which they have been able to give a voice."[39]

This *political* approach for the constitution of the experimental sciences allows for the creation of "a problematic space to be created where one will be able to *attend to* the construction of the difference between science and nonscience,"[40] which "in no way authorizes one to reduce the solutions that are inscribed in it to a common standard."[41] We are a long way from any epistemology that would deliver the "right method" to do science. In each practical case, questions are posed, such as: "Who are the legitimate actors? How are propositions deserving of authority to be selected?"[42]

We have just seen what goes on *in* the laboratory. But we also have to observe what happens outside, something the notion of network could tend to erase. The laboratory creates a "we," a specific collective: "The scientist, as an author, does not address himself to readers but to other authors."[43] (For Stengers, this is importantly different from the creators of *technical* artifacts that have to overcome "challenges that are associated not with the requirements of competent colleagues but with the possibility of reliable performance, endowed with meaning for an essentially heterogeneous collective and related to essentially disparate constraints."[44])

Stengers picks up on what Latour, "in discussing Pasteur, has admirably shown, what is made in a laboratory never really leaves the laboratory."[45] If one closed all the laboratories that thrive on what any such entity makes possible, and so doing make it more stable and indispensable, it would be deprived of its unique ecology and would wither away and die. She goes on: "This is why the researcher, as Bruno Latour has emphasized, is never alone in his laboratory; virtually present are all those whose objections can, and should, be anticipated. On the other hand, absent are all those questions that are excluded by the transplantation [the purification of a phenomenon so the laboratory can study it]."[46] Transplantation is a special "mode of abstraction," she later says, using a Whiteheadian concept. It is not a cognitive operation but a material one, which has to succeed. Transplantation has to succeed keeping the transplanted being alive. Here, transplanted in a new experimental milieu, the phenomenon must "get a voice," that is, the power to *eventually* "confer on the experimenter the power to speak in its name." But transplantation also excludes a whole series of questions in its very operation, questions that can only be asked when one leaves the laboratory. Something one should always do from time to time . . .

But Stengers has another distinction to offer, as Latour was well aware, having followed researchers gallivanting around the world, that while scientific statements only keep their power and effectiveness thanks to the existence and maintenance of laboratories, scientists themselves have numerous reasons for going outside of them. But how? To meet whom? The knowledge coming out of a laboratory is always a "situated knowledge" (one of Haraway's ideas that Stengers has used[47]), a knowledge that is not, in principle, generalizable. She goes on to state what a "civilized" politics would demand:

> Civilized scientists will be the first to affirm that both the reliability of their results and the competence of their objecting colleagues are relative to experimentally purified, well-controlled laboratory experiments, which require ignoring what may be important factors outside the laboratory. They thus acknowledge that whatever they achieve may well lose this specific reliability when it leaves the network of research laboratories.[48]

Here we can distinguish three discursive regimes that the scientists currently use. When they address colleagues, when they speak to *actively interested* funding bodies (in the context of the new "knowledge economy," which might sometimes be more accurately called a "'promise economy' . . . the speculative economy, the bubble and crash economy, [which] has succeeded in recruiting scientific knowledge production"[49]), and when the same researchers present their findings in public, assimilating their work in the Progress that all of humanity will benefit from. This last regime is what politicians so often expect of them: a bit of pedagogy and a prescription for the way forward. All the public is expected to do is *join up*. And lying to them can be justified: "if we tell 'them' about science 'as it is made,' they will lose trust, and if they lose their trust in us, nothing will protect them from irrationality. When talking to children, one must know how to lie or to embellish, and adults must never, ever, argue in front of them!"[50]

This relates to the idea of *obligations* and *requirements*:

> What remains undetermined, even for the practitioners, is the question of how their obligations will be formulated, expressed, "represented," that is to say, the way in which the practitioners justify themselves, define themselves in relation to others. It is not a reflexive question, "What are my obligations?"; it is an ecological one, in the sense that the response also depends on others, on the way in which they require one to think, or not. It actually depends

Sociology or Politics? 59

on a distinction Bruno Latour made in *The Politics of Nature* between that which is of an order of habit, that can be modified, and that which, after testing, must be recognized as an "essential requirement" of a practice, something which cannot be abandoned without destroying the practice itself.[51]

When Stengers speaks of "others," she is no longer talking about "competent colleagues," nor about their actively interested allies, but about those the scientists are asking to join up. Something new is irrupting here. The public could well be in the process of changing. It has learned (and especially at the moment when the Chernobyl cloud was supposed to "have stopped at the French border"!), and now they might be ready to show their recalcitrance. Scientific networks are extending ever further, recruiting new partners, but the researchers do not know too well what to do with some unexpected actors that are not welcome (what do they think they are doing here?) as they become able to challenge the usual discourse that would send them back to irrationality (in part, thanks to the "betrayal" of some of their own researchers). For Stengers, this is the signature of the genetically modified organism (GMO) event.

She returns a number of times to the way in which GMOs were for her

> a [crucial] learning experience [which] marks a before and an after ... What made for an event ... was the discrepancy that was created between the position of those who were in the process of producing more and more concrete, more and more significant, knowledges, and the knowledge of those responsible for public order, [who were] incapable of "reconciling" opinion with what for them was merely a new agricultural mode of production that illustrated how fruitful the relationship between science and innovation was. Even the scientific establishment ... was shaken up. A terrible moment for French science ... The Commission for Biomolecular Engineering ... gradually started to admit that a flow of genes that induce resistance to herbicides was going to be brought about and could cause a problem ... the politicians understood that the situation was out of their control: the scientists were openly divided, public research called seriously into question, militant actions had begun ... but what the politicians had not foreseen is that more than ten years later, they still would not have succeeded in "calming this down" ... What originally engaged me personally was the ignorant

arrogance with which scientists announced a "finally scientific" response to world hunger . . . if the GMO crops affair was an event, it is because there was an effective apprenticeship, producing questions that made both scientific experts and State officials stutter . . . the GMO event constitutes an exemplary case for the bringing into politics of what was supposed to transcend it: progress resulting from the irresistible advances in science and technology.[52]

5
The *Factish* Gods

Let's go back to the way Stengers characterized the sciences by way of a threefold power (the *power* that can be given to things; the *power* that can be given to the experimenter; the *power* to speak in their name). This only refers to the theoretical-experimental procedure taking place in classical laboratories. This procedure is not a right, it relates to "requirements" whose fulfillment does not depend on the researchers: the object has to be *transplantable*, not just *referred*. As Ginzburg had argued, sciences are differentiated. How can we understand sciences like "geology, evolutionary biology, climatology, meteorology, and ecoethology, as sciences that address situations that cannot, as such, be 'purified,' reduced to laboratory conditions."[1] How can one address entities that lose all of their attributes if they are *purified* and *abstracted* in order to enter something that looks vaguely like a laboratory, and that will only deliver a boring and sterile ("monotonous," Stengers often says; "boring" adds Latour) "postage stamp collection"?

Pluralizing the ways of doing science – and at the same time characterizing theoretical-experimental success in such a way that it cannot be the endlessly reproducible model – Stengers (unlike Ginzburg) advances a new kind of protagonist who construct "models" rather than theories. These models present themselves as always incomplete fictions. They are not looking to purify their objects, as in the cases of Galileo and Pasteur; on the contrary, every new detail that can be integrated into the model is welcome. The model articulates what she calls, in reference to Deleuze and Guattari, a "*disparate* multiplicity." Mathematics, here, enjoys "privileged ties with the speculative powers of the imagination, and not with a 'theoretical truth' of the world."[2] Models need data-based tools as simulation devices. In this way, the experimental, the descriptive, the explanatory communicate "in a new, fictional mode."[3] So we are dealing with models that have another way of *getting a grip* on things, a new "mode of abstraction." We can, for example, see here how many climate specialists work.

Now a new political problem emerges: "How should one regulate the relations between the denizens of the two types of laboratory, who are on vectors of divergent modes of engagement?"[4] Classical experimental science now finds itself confronted by other practices which are *just as inventive and risky as its own*, but "by their very existence put in question the power of truth that defines this enterprise."[5]

Having dealt with models, Stengers goes on to the example of the descendants of Darwin, at the origin of another way of doing science: "For a long time, 'Darwinian' science has been presented in a form that enabled it to claim the same power to judge as the laboratory sciences. Natural selection had to be all powerful so that its representative could claim the power to judge, to explain, even if rhetorically, the history of living creatures."[6] But it is an unavoidable characteristic of "fieldwork" that it is impossible to stabilize responses that are the same everywhere. The right question is, first of all, "how to 'describe,' not "how to 'inter-relate.'"[7] We had a glimpse of this in Latour's account of the Boa Vista inquiry. When it comes to evolution, general concepts like adaptation, survival of the fittest, etc., have proved "to be empty of a priori explanatory power: simple words coming to comment on a history after it has been reconstructed."[8] She often comes back to this: "selective explanation was doomed to substitute one and the same monotonous, 'epicyclic' answer (in this case, 'adaptationist') for the multiple questions that arise for those who study living beings."[9] She says this, wryly, in reference to the attempt to save the rule of the circle in pre-Keplerian astronomy. So biology appeals to another "mode of abstraction," which is different from the one needed in the theoretical-experimental laboratory, or the one needed to construct models.

The *Factish* Gods 63

In such a situation, scientists lose the power to purify or to judge. They "have to learn how to tell stories."[10] They are investigators, as in the portrait sketched by Ginzburg in his famous essay. They are hunters, he wrote, like Sherlock Holmes:

> Man has been a hunter for thousands of years. In the course of countless chases, he learned to reconstruct the shapes and movements of his invisible prey from tracks on the ground, broken branches, excrement, tufts of hair, entangled feathers, stagnating odors. He learned to sniff out, record, interpret, and classify such infinitesimal traces as trails of spittle. He learned how to execute complex mental operations with lightning speed, in the depth of a forest or in a prairie with its hidden dangers.[11]

This is why these kinds of researchers, that Stengers will call "historians of the Earth"[12] (from the point of view of Latour, it could be considered as a premonition), could be of interest to model makers. They distance model makers from the ideal that is the theoretical-experimental sciences, that is to say, from the rare cases of success that clarify functionalities and orient them toward stories . . .

The relations enjoyed by so-called "field" scientists with each other have nothing to do with those who create the community of laboratory experimenters: their observations *can be contradicted without being invalidated*. No one knows a priori how selection works, nor what is due to it. Stengers thinks Stephen Jay Gould's 1989 *Wonderful Life* deserves as much praise as Galileo's *Dialogue*: "As Stephen J. Gould has admirably demonstrated, what gives the evolutionary sciences their robust character is not the 'proof,' but the number and variety of cases that become intelligible and interesting in a Darwinian perspective."[13] Today, she could build on a new generation of researchers who are influenced by Lynn Margulis but lay claim also to Darwin. This is how Carla Hustak and Natasha Myers express it:

> By leaning into Darwin's experiments, hovering with insects as they lap at floral nectaries and feeling ourselves pulled by the labellum's lure and the tobacco plant's volatile plumes, we can begin to tell new kinds of stories. As we hitch a ride on this involutionary momentum and draw attention to the rhythms of these intimate relations, we can disturb the militarized and economic logics that pervade the sciences of ecology.[14]

Uncertainty is the hallmark of the field sciences. As we have seen, the threefold power that characterized the production of laboratory facts is truly an exception.

But Stengers proposes another "type of variation," a "mode of abstraction" which she links to Vinciane Despret's work[15]: sciences of contemporaneity in which the "production of knowledge" is inseparable from the "production of existence."[16] This concerns all sciences dealing with living things that *"no means can render indifferent to the fact that they are interrogated."*[17] Despret, Stengers, and Latour will turn the *recalcitrance* of these living beings, their capacity to differentiate "good" from "wrong" questions, into a requirement that contrasts with the "indifference" of the actants to the questioning apparatus of experimenters in the theoretical-experimental laboratory. This indifference means that "how they are to be addressed will have to be invented."[18] So, in 1997, Latour introduces this new concept:

> Stengers has suggested applying a method to the human sciences that can change its style profoundly . . . Stengers' view is that there is a direct relation among the quality of a scientific discipline, the interest of an experiment or a theory, the risk run by the researcher and the recalcitrance of the objects. Knowledge is no longer measured by the yardstick of its objectification, but by the risks shared by the observer and the observed.[19]

Stengers comes back to this on the occasion when scientists were complaining about the way sociologists of science were planning to study them. She jokes:

> This is why Bruno Latour, apropos the social sciences, suggested calling a mistake that has a fortunate consequence, a *felix culpa*, the fact that certain practitioners of experimental science, once they understood the meaning of the questions critical social scientists were asking of their practice, protested violently . . . For Latour, the . . . social sciences should take the lesson on board: they were at fault not only in this case, but also every time the people they study reply "without making a fuss" . . . Only with "recalcitrant" protagonists . . . can a relation be created that has a claim to scientific value.[20]

This is also the price to pay for them to be adventures, and not the revelation of a hidden rationality, which is what Stengers had already learnt about hypnosis with Chertok.

Stengers will introduce a new actor into the debate: Tobie Nathan. She has a talent for detecting the most promising talent in the most diverse fields, in this case ethnopsychiatry. She was well prepared for this encounter because of her approach to psychiatry when she worked with Chertok.

It is from Nathan that she learns that the techniques of healers are "robust," not because they can be linked to any very general "symbolic effectiveness," as Lévi-Strauss thought, for example,[21] which "would designate 'westerners' as those who judge and put all the others to the test" but because "what they implement is capable, from generation to generation, of fabricating technicians, technicians whose practice is the object of the competent, discussed, and well-informed attention of the patient and their family, who know very well that all healers are not of equal worth."[22] And, she adds, "I want to point out a crucial element in Nathan's writing on 'nonmodern' therapeutic practices: it is as if it were imperative for these therapeutic techniques to confer upon the illness a significant value based on a strategy that creates obligations where modern practices have requirements."[23] And she cites this passage from Nathan:

> [reading] coffee grinds provides a considerable methodological advantage over the Rorschach test. Naturally, it is also a kind of projective test, but one undergone by the clinician – who is obviously, in this case, a *seer* – rather than by the subject. And since it is administered in the presence of the subject or an object representing the subject . . . the reading of the coffee grinds can only provide information about the state of the relationship between clinician and subject, not about the hidden hypothetical nature of the so-called subject. So, the reading of coffee grinds appears, strictly speaking, to be a technical procedure intended to force the clinician to speak only about the interaction that he has established with a person, and consequently to produce usable clinical material.[24]

Nathan comes back to this a number of times: "I now consider that the only object of a *truly scientific psychopathology* should be the most detailed description of *therapists and therapeutic techniques* – never of patients."[25]

If this practice interests Stengers so much, it is because it allows for the characterization of "the event, where a science, 'psychology,' would leave the paths of method and its rough 'one-size-fits-all' explanations for another, where what is discovered is the obligation to entertain the possibility of being cast into an adventure by whatever one is dealing with."[26] So it is not just immigrants who can benefit from what ethnopsychiatry brings:

> And here Tobie Nathan's challenging question comes into play: wouldn't we have something to learn from those healers, whose common characteristic is that they are not haunted by the ideal of a Royal Way endowed with the capacity to disqualify others, but rather by having cultivated what one could call, following Nathan, an art of influence. It is appropriate here to distinguish "influence," in Nathan's sense, from "suggestion," "imagination," or "placebo effect". . . Suggestion is what we are all likely to be able to make use of, like Monsieur Jourdain, without even knowing it. Influence implies the expert; it implies a knowledge whose power and interest are, as Nathan shows, to "technologize the therapeutic relation."[27]

It was important to make this specific because, with a kind of *us too*, modern doctors carry on their war against traditional healers (which began well before medicine could be said to be scientific) and have been a little hasty in appropriating the influence: the placebo effect that they had reconfigured as a "lab coat effect," allowing them to double dip (they have on their side both science *and* the effects of the imagination, henceforth rendered banal, while being totally ignorant about them). The placebo effect, like suggestibility, is a "tote bag" idea: "What was problematic has been eliminated, absorbed in a more general category, so general that it poses no problem and can be forgotten after having done its elimination duty."[28]

But for Stengers the question of the Great Divide became decolonial: "Nathan . . . denounce[s] not only the way in which we 'care for'" migrants and their descendants but also the way in which we, with the help of our standards, facts, and good intentions, deny their essential right to maintain the obligations and requirements of their culture."[29]

For his part, Nathan could only go along with Latour's problematization of the Great Divide: "Following Bruno Latour's reflections, we have abandoned the great divide 'between those who believe and those who know'. . . breaking with a tradition, which, on the topic of mental health, has hierarchized knowledges, placing 'scientists' on one side and 'all the others' on the other, stuffing 'beliefs and traditional representations' into the same bag."[30] Stengers writes: "As Bruno Latour has often written, our true difficulty is that 'we believe in belief,' and that when we are dealing with something whose power we have to learn how to characterize, we 'psychologize' the problem."[31]

Stengers reminds us that some anthropologists have been able to turn "the opposition – they who believe versus we who know – into a professional flaw. They have also refused any passport permitting them to feel at home everywhere, or to assert that 'nothing that is human is foreign to us.' What Eduardo Viveiros de Castro characterizes as the task of 'decolonizing thought'..."[32]

In turn, Latour takes an interest in Nathan's work, spending more than a year at the Georges Devereux Center, and a little book will be the result, published in 1996 for the first time. It is this experience that will later enable him to propose an entirely separate "mode of existence" (MET) for metamorphosis, which he could just as easily have called (SOR) for sorcery:[33] "In applying to humans an epistemological model which no researcher had ever applied to objects, and thus imitating a nonexistent scientific model, psychiatrists have failed to see the specific originality of the therapy."[34] Psychiatrists would thus try to imitate how they imagine scientists do their work (what some philosophizing physicists might let them believe).

In this little book, he distances himself, albeit for a moment, from Nathan, something Stengers never did: "Must I believe what the ethnopsychiatrists say about what they do, or should I follow their practices? ... What is true for philosophers of the sciences may also be true for ethnopsychiatrists."[35] What interests Latour about Nathan (he will say, "the migrants heal us"[36]) is that he offered him a site (the Georges Devereux Center) "other than a laboratory to work out this notion of subjects, which is the counterpart of the notion of facts."[37] So this was an apparatus as *artificial* as a laboratory tool designed to "depsychologize," just as the study of laboratories allowed for "de-epistemologization."

This text is also the occasion for Latour to explain why a term should be banished: *representations*. It has no place either in Latourian anthropology, or in Stengerian philosophy, or with Nathan. It is never necessary to add "representations" to make the world more real ... let alone more enchanting. In this way, ethnopsychiatry does not work on patients' "representations."

Thinking things through in this manner would be to repsychologize them, putting back in their heads exactly what should remain outside. We know that a lot of psychoanalysts, briefly trained in ethnopsychiatry (who have absorbed what would more accurately be called "transcultural psychiatry" in reference to a strand recognized officially in the United States, charged with the task of writing one of the chapters of the famous *DSM* (*The Diagnostic and Statistical Manual of Mental Disorders*), quickly reverted to representations and got busy repsychologizing patients, interpreting and translating what they say into classical Freudian references.[38] Conversely, at the Georges Devereux Center, interpretations are banished! Nor, in the work of Latour, Stengers, or Nathan, is there any place for the symbolic. From 1984, Latour was writing:

> The symbolic is the magic of those who have lost the world. It is the only way they have found to maintain "in addition" to "objective things" the "spiritual atmosphere" without which things would "only" be "natural"... Those who wish to *separate* the "symbolic" fish from its "real" counterpart, the fishy fish and the fishy poison [*le poisson poissonnant et le poison poissonné*] should themselves be separated and confined.[39]

Here Latour links up with Stengers, but again by his own route: there is an error from the start as to what constitutes the human *interior*. He talks about it in 1999:

> I encountered it in Africa without understanding it, some thirty years ago; it was only in Tobie Nathan's practice that I myself recognized the difference between encountering a patient under the auspices of anthropology and encountering a patient under the auspices of risky diplomacy (Nathan 1994). Naturally, those who cry, as did a well-known Parisian psychoanalyst, "Without a universal unconscious, there is no longer a French Republic" accuse Nathan of culturalism – just as the Sokalists accuse me of social constructivism.[40]

As soon as Nathan is the object of nasty critique, often coming from psychoanalysis, accusing him of racialism (which is also what the sociologist Didier Fassin does[41]) and of rejecting universal republicanism, Latour and Stengers together construct a platform for his defense:

The *Factish* Gods

It is not just the immigrants being treated at the Georges Devereux Center, it is us, this ailing France, much less under threat from the "rise of irrationality" than by the absence of a sorting mechanism for learning from others, all the others, those coming "from below" as well as those "up above". Relegating Tobie Nathan to the shadows of the archaic means losing the opportunity to benefit from what migrants have to offer for the redefinition of the Republic before it "loses its soul" in the slow disintegration toward a Europe that is itself defined by the necessary but difficult conditions of that universal that no one, it seems, has a need to think about: the global market.[42]

Rather than a precise and detailed analysis of healing such as is carried out at the Georges Devereux Center – it's only in the second part of *On the Cult of the Factish Gods* that he deals succinctly with this – this book is the occasion for Latour to deal with the inconsistencies of *critical thought*, of the idea of belief, and to come up with a new, groundbreaking idea to solve the problem he had been working on ever since *Laboratory Life*, to find the right words for saying that a "fact" [*fait*] is "made" [*fait*] – following the etymology – because it is "well made": *factishes* [*faitiches*]. This is the neologism he made out of the association of "facts" and "fetishes." The question "Is it well made?" will become a transversal one, the key question of the *Inquiry into Modes of Existence*. It allows Latour to address not only fetishes and scientific facts but also legal judgment and even religious speech. In each case, what matters is to decipher what "well made" demands.

> [T]hose who are accused, most wrongly, of naively believing in their fetishes have in fact preserved this precious *savoir faire*: they can say in the same breath and without trembling that they made with their own hands the thing that nonetheless saves them and gives them life, that possesses them and holds them. . . . can I now forget the comminatory question, "Is it real or is it made?", and here, too, substitute the question: "How do you tell the difference between what is *well* made and what is *badly* made?"[43]

Latour warned us in a text published a few years later:

To think in terms of "factish" requires some getting used to, but once the initial surprise at such an outlandish twin-horned form passes, one begins to regard those obsolete figures of object and subject, the made and maker, the acted upon and the actor, as more and more improbable. I shall not attempt to transcend them once again, through the dizzying effects of dialectics, but instead I will simply ignore them, signaling in passing their complete irrelevance.[44]

He comes back to the way in which critical sociology plays a double game in which it always comes out the winner:

The critical thinker triumphs twice over the consummate naivety of the ordinary actor, seeing the invisible work that the actor is projecting onto the divinities who manipulate him, but also seeing the invisible forces that drive the actor, who believes that he manipulates freely! (Critical thinkers, offspring of the Enlightenment, ceaselessly manipulate invisible things themselves, as we can see; the great liberators from alienation produce endless numbers of aliens).[45]

And he scoffs, "No emperor is more exposed in his nakedness than the critical sociologist who sees himself as the only lucid person in an insane asylum."[46] Stengers says the same thing:

But the fact that today critique can end up in the sadness of "it's just a social construction" marks the end of the intensely "constructivist" moment... Perhaps one might say that critique, which certainly was a remedy, has become a poison because it has not known how to defend the actual truth of what is constructed, of what succeeds in holding together and making hold together that which is fabricated, and yet has the power of a "cause," which makes those who fabricated it think, act, and feel.[47]

In *Irreductions*, Latour had already written: "Those who talk of metalanguage must mean, I think, the pidgin of the masters which is too impoverished even to translate what is said in the kitchen."[48] He upholds a sociology (which he calls that of the actors themselves [*les zacteurzeumêmes*]) in which the aim is certainly not to reveal a truth that the actors could not see; the new sociology that he promotes with the Centre de sociologie de l'innovation at l'Ecole des Mines "takes as its job to convey forms of life in common, against the *diktats* of critical thought, from the rear of the shop to the window display."[49] The human sciences only bring "a slight surplus" to what the actors know already. In 1999, he wrote:

The Factish Gods

To assert that underneath legitimate relationships there are forces invisible to the actors, forces that could be discerned only by specialists in the social sciences, amounts to using the same method for the metaphysics of nature as was used for the Cave: it amounts to claiming that there exist primary qualities – society and its power relations – that form the essential furnishings of the social world, and secondary qualities, as deceitful as they are intensely experienced, that cover with their mantle the invisible forces one cannot see without losing heart. If the natural sciences have to be rejected when they employ that dichotomy [the bifurcation of nature], then we have to reject the social sciences all the more vigorously when they apply it to the collective conceived as a society . . . The social sciences . . . have a much more useful role than that of defining, in the actors' place and most often against them, the forces that manipulate them without their knowledge. The actors do not know what they are doing, *still less* the sociologists.[50]

Later, we will have more to say on this distinction between primary and secondary qualities ("real nature'" and "apparent nature"), a distinction that goes back to the Subject–Object opposition we started with in the introduction, and which then took on the name of the "bifurcation of nature" (see chapter 7).

Latour treats this in detail in *Reassembling the Social*, first published in English in 2005, while Stengers makes a similar distinction between *unveiling* and *characterizing*:

> To unveil would be to have one's heart set on passing from perplexity to the knowledge that, beyond appearances, judges. On the other hand, to characterize, that is to say, to pose the question of "characters," is to envisage [a] situation in a pragmatic way: at one and the same time to start out from what we think can be known but without giving to this knowledge the power of a definition.[51]

She extends this point of view by broadening it to include alienation, an important concept for some Marxists:

> That is also why we don't like theories of alienation, haunted by the fact that "people" seem incapable of "becoming conscious" of the truth of their situation. Those who pose the problem in this way are supposed to know a bit about this truth: they don't need people to think; rather, they regret that the truth that they possess, and which should by rights be of value for the people, doesn't enlighten them.[52]

More Great Divide mischief!

In order to challenge the Great Divide, the idea of factish fuses two terms: "fact," which seems to relate to external reality; and "fetish," the subject's mad beliefs. Crucially, the factish is of the order of the event, and not of the Truth. It is the radical solution "that allows one to abandon critical thought, forget notions of belief, magic, hypocrisy, and autonomy, losing the stunning mastery that has made us Moderns and proud of it."[53] All those who establish/construct facts are "slightly left behind" by what they have constructed. "There have never been any Barbarians; we have never been Modern."[54] No one has perfect mastery of what they have made. One always has to "trick" in experimental laboratories as much as in practices of healing.[55] It is not the question of "truth" in its generality which is at stake here. If anthropologists were often mistaken about what was going on at the Georges Devereux Center, it was because they were "looking for an authenticity of ethnicity that they [couldn't] find, and they fail[ed] to see that the originality [of the Center] stems precisely from its artificiality."[56]

Stengers came up against the same argument when she became interested in neo-pagan witches, in particular Starhawk:

> If neo-pagan witches could be identified with a "true," authentic tradition, the manner of their resistance could be respected because we have the habit of tolerating the survival of traditions, indeed even of respecting the wisdom immanent to them. The test stems from the experimental, "fabricated" character of their rituals and the undecidability that they confront us with . . . What makes people uncomfortable, what is difficult to accept, is that witches are pragmatic, radically pragmatic: truly experimental technicians, experimenting with effects and consequences.[57]

Once again, this does not mean that all is possible. On the contrary; in ethnopsychiatry, the patient or the family undergoing treatment are also situated as *experts* of their situation (an expertise reinforced by a member of the team who shares their language and culture). It is harder to imagine a stronger constraint.

Latour uses Pasteur as an example once again:

> We do not want to drown Pasteur, with his attentiveness to the precise gestures that revealed his ferment, any more than we wish to

The *Factish* Gods

lose our Candomblé initiate [a cult introduced into Brazil by African slaves] who is fabricating his own divinity . . . Pasteur does ask us to recognize that the ferment has all the autonomy of which it is capable. The Candomblé initiates in no way claim that their divinity speaks to them directly through some heavenly voice since they admit, just as ingenuously, that their divinity is at risk – for want of "know-how" on their part – of becoming an "endangered species."[58]

Obviously, it is not a case of saying it is the same thing. "Each requires its own particular forms of existence for which lists of specifications have to be drawn up," which prefigures the *Inquiry into Modes of Existence*. In both cases, "artifice is the friend and not the enemy of reality, whether in terms of the laboratory setup or the creation of ethnic affiliations."[59]

Stengers endorses the idea of factishes: "It is in order to allow for a more 'civilized' encounter that Bruno Latour introduced the term 'factish,' intended to prevent the critical visitor from characterizing what laboratories bring into existence in terms of 'belief.'"[60] That is, she emphasizes that Latour is not only speaking about the Moderns but is distinguishing himself clearly from the sociologists of science for whom everything is a social construction. Latour overtly recognizes this clarification when he writes, "I had indeed detected the phenomenon, but it took me twenty years to understand the synonymy of the two verbs 'to build' and 'to overtake' [*construire et dépasser*]."[61] And, he adds, "Social constructivism is the creationism of the poor in spirit."[62] In 2012, he will remember: "I had no way of knowing that it would take me a quarter of a century to get myself out of the misunderstanding created by the use of the word 'social' and from all the complications that turned out to be attached to it, to my great surprise."[63] And he will recognize the important role Stengers has played in distancing him definitively from the Anglo version of STS. He puts it bluntly: "It is to Stengers, whom I had known since 1978, that I owe the constant disruptions that she imposed on all the social explanations – even those improved by the ANT."[64]

She herself will then take on the idea of factish, which conveys well how "facts" are both fabricated *and* independent of the tests that have made them appear if and when they have been well fabricated.[65] But she will have fun multiplying the number of varieties of them in her continual effort to distinguish different practices among them (what requirements? what obligations?), especially in *Cosmopolitics*. Thus there are factishes that are experimental, physico-mathematical, enigmatic, divinatory, problematic, and promising . . . a way of accepting Latour's idea and making it blossom.

According to Latour, the irruption of new entities (like factishes) change society, especially thanks to the laboratory. A society with germs is not the same thing as a society that is unaware of them. So it is not the case that an epidemic comes along, as many say, to "reveal" what a society is; it changes it completely by welcoming new actors. The Romans leant this the hard way![66]

Factishes will only outrage those for whom scientific statements are merely "social constructions": "We may well speak of men, societies, culture, and objects. But everywhere there are crowds of other agents that act, pursue aims unknown to us, and use us to prosper."[67] Finally, relativists and rationalists (critical sociology) have something in common that is poisoning them. For both of them, society is defined in advance and only takes human actors into account. Latour keeps coming back to the Pasteur example: "Hygienists also started with a fixed state of Society – the class struggle – and a determined state of Nature – the miasmatic diseases. When Pasteurians offered them the microbes, this was a new and unpredictable definition both of Nature and of Society: a new social link, the microbe, tied men and animals together, and tied them differently."[68] Later, he drew a more general lesson from this: "ANT simply doesn't take it as its job to stabilize the social on behalf of the people it studies; such a duty is to be left entirely to the 'actors themselves.'"[69]

This is something Latour emphasizes throughout his *oeuvre*: nothing can be explained via "the social"; it is what remains to be explained because it always comes later, as a result. One very clear account of this is given in the *Inquiry into Modes of Existence*: "'The social' . . . does not define a material different from the rest, but rather a *weaving* of threads whose origins are necessarily varied. Thus, in this inquiry, 'the social' is the concatenation of all the modes."[70]

It will be up to Stengers to come back to the question of techniques: the technical objects that Latour speaks of, and Nathan's therapeutic techniques. In 2002, after her long labors on and with Whitehead, who had for a long time taken up a lot of space in her thought,[71] Stengers returns to the question of "techniques," firstly those that are active in therapeutic fields. And it is on the question of techniques and technicians that she will advance that "Tobie Nathan's ethnopsychiatry and Bruno Latour's 'politics of nature' enter into a close relationship."[72] If it is indeed a "close relationship," we should pause on it for a moment. She begins with the idea of techniques as defined by Latour and correlates it with a political question:

> I will approach the delicate question of an evaluation of therapeutic techniques. In a general manner, let us say that it is posed in terms of the interrogation enunciated by Bruno Latour in his *Politics of Nature*: do we wish, or do we accept, to live with beings produced by this technique? Let us stress that this question is always concrete: a GMO created by Monsanto, for example, is not the same being as the possible GMOs that might one day be created by research led in a concerted way with concerned Indian peasants.[73]

Techniques, indeed, create new beings. They are transformative; they are essentially different from the sciences. Stengers writes:

> We have passed from a "scientific" question (can such and such a phenomenon be exhibited as "a function of," or as "allowing itself to be explained by") to a question that designates first of all the successful production of a new being . . . we have changed our practical domain. For the construction of knowledges that aim at such productions, that seek to stabilize them, belongs to "technicians," not to "scientists." Unlike experimental practices, technical practices escape all opposition between fact and artifact, reliable witnesses and compliant witnesses . . . No one would think of accusing a technical apparatus of "fabricating" what it produces . . . Correlatively, while the notion of "science" in the modern sense cannot, or should not, be generalized to modes of knowledge cultivated by other traditions, there is no reason to restrict the notion of technique to our "modern" practices.[74]

In her little book *L'Hypnose entre science et magie*, Stengers will therefore take up again the question of psychoanalysis, more than ten years after the lecture she gave to the crowd of psychoanalysts that we presented in chapter 2.

The question of techniques is always a concrete one. Nathan, in fact, always refused terms like "trance" about the different traditional therapeutic techniques, which suggests "going 'beyond,' toward what would explain them 'for us,'" which consists of bringing all the very diverse therapeutic *techniques* back to the monotony of the same: we know it is only this . . . The trance is an *abstract* idea, one incapable of coming to terms with the concrete. Let us remember that here *abstract* means purified, deprived of its environment – something scientists have very good reasons to do in the experimental laboratory, but which, once out of the laboratory, becomes merely mimetic, being the hallmark of practices that can be called "modernist," that are incapable of inventing the requirements and the obligations that would render them fertile. Nathan certainly has issues with the "lack of resistance characteristic of the modern epoch to the intolerant rule of abstractions that declare everything that escapes them frivolous, insignificant, or sentimental."[75] Stengers writes:

> from the point of view of ethnopsychiatry . . . even the fact of speaking of "trance techniques" is somewhat dangerous, for the term "trance" invites us to think about what happens to humans "in a trance" and not about what the technician is addressing. To pose the question of trances, or of the trance, is thus to pose as judges, claiming to know the practice of technicians better than they do themselves, without even having to encounter them. We would not have to stop at the fact that, for them, what we call trance is first of all what connects with invisible beings and allows them to be negotiated with. We would have the right to separate their technique from any kind of thought; that is to say, to think in their place . . . Ethnopsychiatry, on the other hand, tries to take seriously the techniques of the practitioners of healing, and does so by guarding against going "beyond," toward what would explain them "for everyone."[76]

She adds: "For all of these practitioners [healers], the trance as such is not interesting . . . Other speculative philosophers have created other words. Hence for Deleuze and Guattari, a key term is 'force,' and every 'assemblage' must be understood starting from the forces that it captures, or can make felt or perceived."[77]

Because if we have to characterize these beings that the healers summon, with whom they have dealings, it is through their effectiveness (a term to be distinguished from symbolic effectiveness!). Here, too, explains Stengers, we "come across the manner in which Bruno Latour has characterized our curious passion for anti-fetishism. The fear of anthropomorphism, the threat of being convicted of interpreting instead of sticking to the severity of bare facts, marks us as 'anti-fetishism addicts,' addicted to the criticism of those we accuse of attributing an autonomous existence to what they, in fact, fabricate."[78] But what is the right way to ask this effectiveness question? Tobie Nathan taught us (during a treatment consultation for former members of sects) not to take the *weakness* of patients as a starting point, which would inevitably send us back to psychology, but the *force* of an apparatus, what the technicians succeed in "selecting, convoking, stabilizing, in short, cultivating."[79] What is the ethology of the beings summoned by healers? Nathan discusses this in detail in a text appropriately entitled "Dealing with Devils."[80] Dealing, or trading, with these beings is indeed what technicians of healing must do and it means obtaining their willingness to enter into a "commerce" relationship. According to Stengers, the ethnopsychiatric invention entailed that "to take seriously the efficacy of [such] a technique demands understanding it as addressed to something more powerful than the technician."[81]

So now we can come back to psychoanalysis. Freud's masterstroke was also a power grab. At the very moment when he made his break from hypnosis and was looking to invent a practice that was at last scientific (this is the theme of the third narcissistic wound), he refused what is the main requirement for a science:

> If modern sciences merit the designation of adventure, it is because, as long as a controversy is open, everyone involved is required to recognize the "competence" of adversaries and the legitimacy of their skepticism. The primordial measure of the claim to truth, in the scientific sense of the word, is its ability to overcome all competent objections. In contrast, this ability to overcome all objections is precisely what is denied by the Freudian argument.[82]

This is what psychoanalysts following the path opened by Freud refused to do: any objection is analyzed as a "resistance" to psychoanalysis: "the identification of critique with a symptomatic resistance applies as much to dissident analysts as to external critics."[83] In other words, it can be said that, like technicians of healing, analysts associate what they confront with a power that one has to recognize in order to address it.

But a second point has to be noted here. In wanting nevertheless to turn psychoanalysis into a scientific practice, Freud invented "a being of a new kind,"[84] the psychoanalytic unconscious. If the Freudian unconscious must, *like any other scientific entity*, resist the accusation of being a "mere" interpretative tool enhancing the suggestive influence of the therapist, then one has to accept paying the price: the unconscious is defined as indifferent to suggestion, that is, indeed, "inaccessible to civilized commerce."[85] The consequences are devastating:

> [This] is perhaps the drama of Freud . . . not having been able to use the interesting idea of "commerce" . . . And doubtless one of the reasons for the fascinating character of the Lacanian unconscious is that it has inspired the most radical and insulting slogan with regard to all commerce . . . Correlatively, the link between analysis and "cure," already largely speculative for Freud (how can consciousness raising be a cure?), became an object of derision. . . . people who fearlessly frequent beings liable to invade a life, to devour a soul . . . have, I understand, a name: they are sorcerers . . . can the Freudian–Lacanian unconscious, indifferent and thus theoretically inaccessible to the civilized compromises that other "supernatural" beings (sometimes) accept, participate in anything other than practices that bewitch, capture souls thenceforth destined to work in its service? Analysis fabricates analysts, Lacan wrote. A fine definition for a sorcerer's machine because the sorcerer only knows how to create sorcerers.[86]

"Ekudi: I think your way of going about things is very dangerous, *Fofo*; and especially, it can't be of much good . . . Your patients must remain sick for a very long time . . . years or maybe even decades . . . have you been able to cure a single patient in this way?"[87]

6
The Parliament of Things: Doing Ecology

In *We Have Never Been Modern* (1991), Latour drew up a general picture of the ambitions of the Moderns that he called their "Constitution" – which is obviously implicit and so needs to be clarified. *We Have Never Been Modern* comes as an echo of *Irreductions*, and thus extends a cycle which will lead to *Inquiry into Modes of Existence* in 2012. But by 1999, *Politics of Nature* inaugurates another theme, that of ecology. It opens with a sentence that reveals its amazing ambition: "What is to be done with political ecology? Nothing. What is to be done? Political ecology!"[1] And his conclusion is, "Can you really say, without blushing, still believing it, that the future of the planet consists in a melting away of all cultural differences, in the hope that they will gradually be replaced by a single nature known to universal Science?"[2]

It is in *We Have Never Been Modern* that Latour uses the example of the controversy between Hobbes and Boyle, distancing himself from the interpretation given by Steven Shapin and Simon Schaffer in a book already discussed earlier.[3] For Latour, what matters is that with Boyle a new kind of witness appears: "Here in Boyle's text we witness the intervention of a new actor recognized by the new, Modern, Constitution: inert bodies, incapable of will and bias but capable of showing, signing, writing, and scribbling on laboratory instruments before trustworthy witnesses. These nonhumans . . . are even more reliable than ordinary mortals."[4] And it is in the same movement that the Moderns' constitution with two chambers is invented. It separates, dissociates, humans and nonhumans, science and politics, Nature and Society, facts and values; the Great Divide once again.

The word "representation" means two things that really have nothing to do with the idea of mental representation but relate more to public life: on the one hand, the political representation of humans and, on the other, the scientific representation of things: "The political spokespersons come to represent the quarrelsome and calculating multitude of citizens; the scientific spokespersons come to represent the mute and material multitude of objects. The former translate their principals, who cannot all speak at once; the latter translate their constituents, who are mute from birth."[5] It will therefore be necessary to take seriously this *representation* word, which is no more a metaphor in the case of the sciences than it is in politics.

But the way the Moderns speak of themselves like this does not correspond to what they do! They are constantly mixing up humans and nonhumans, science and politics. It is precisely the function of researchers to introduce always new nonhumans into society. The Moderns produce knowledges and invent technologies (what Latour will call *hybrids*) that change the way so-called "social" problems are formulated. Pasteur is the best example of this, as perfectly expressed in his time by the bacteriologist Albert Calmette (1863–1928), the inventor of the BCG (tuberculosis) vaccine: "Of course we should try to eliminate overcrowding, insalubrious dwellings, alcoholism, and misery, which are, quite rightly, considered to be the principal causes for the weakening of the organism . . . But what use would all these measures be if we do not dry up the sources of evil, and if we do not first remove the bacillus?"[6]

It's a tight situation: "From now on there will thus be two different histories: one dealing with universal and necessary things that have always been present, lacking any historicity but that of total revolutions or epistemological breaks; the other focusing on the more or less contingent or more or less durable agitation of poor human beings detached from things."[7] In such a situation, ecology can only be paralyzed, impossible to think about. What can make it good to think with?

The Moderns were able to set themselves up as the only civilization that knew how to separate Nature and Culture, while in fact they have never stopped doing the opposite. They thus brought about a separation from the poor old pre-Moderns, who, for their part, confused these things: "We westerners cannot be one culture among others, since we also mobilize Nature. We do not mobilize an image or a symbolic representation of Nature, the way the other societies do, but Nature as it is, or at least as it is known to the sciences."[8] What is to be done? "Symmetrical anthropology needs a complete overhaul . . . so that it can get around both Divides at once by believing neither in the radical distinction between humans and nonhumans at home, nor in the total overlap of knowledge and society elsewhere . . . [it] refrains from making any a priori declarations as to what might distinguish westerners from Others."[9]

The best way of differentiating scientific collectives is by their size. "Even though they might be similar in the principle of their co-production, collectives may differ in size . . . Relativists, who strive to put all cultures on an equal footing by viewing all of them as equally arbitrary codings of a natural world whose production is unexplained, do not succeed in respecting the efforts collectives make to dominate one another."[10] And, he adds, "[The relativists] neglect even more thoroughly the enormous efforts westerners have made to 'take the measure' of other peoples, to 'size them up' by rendering them commensurable and by creating measuring standards that did not exist before – via military and scientific expeditions . . . Why do we like to transform small differences in scale among collectives into huge dramas?"[11] Thus we should break with the old-style modernization that "can no longer absorb either other peoples or Nature."[12]

Stengers totally disagrees with the idea that "huge dramas" may be reduced to (mere) differences of scale. It is one of the few times she says no! For her, the idea of *absorbing* others is a subtle but insistent way of not giving up on the Great Divide, and to keep us in the position of judges:

> What poses the problem is the way in which extraction [in the sense that the laboratory *extracts* and *abstracts* from a situation that which becomes its sole object of possible study] and modernization have been linked, transforming the question "What can we learn here?" into a principle of judgment that identifies what has

been extracted with what really matters, and relegates the rest to an overlay of beliefs and parasitical habits. A genuine prohibition is needed to dissolve this link: *no one should be authorized to define generally "what really matters."*[13]

If this prohibition is not respected, the Great Divide, having left by the door, comes back through the window. Knowledges with long networks creating commensurability should be thought about as extending their definition of what really matters. Certain lengthenings are caricatural (one can, for example, think of the psychiatrists' *DSM*, whose network is particularly extended). Others should be stopped, as we saw in the case of the GMOs. In such cases, we should think "in terms of connivance" what "seems to bring together merchant, then capitalist practices, and scientific practices." Because, she insists, between them "there is not a hidden identity, which would transform their complicity into a destiny, but a convergence relative to interests, posing a political problem that can receive very different solutions . . . For everything changes when one leaves the laboratory."[14] Stengers imagines scientists who would know "that the style which belongs to the risks of the test – the invention of ways of purifying a situation in order to make it constitute a reliable witness – changes meaning when it is a question of choices bearing on irremediably concrete situations, where words, if one is not careful, have the power to disqualify, to silence."[15] There are, she says,

> researchers–authors–critics for whom "leaving the laboratory" deserves to be thought through with the same scrutiny as occurs inside the laboratory. Let's just say that they are not only in the minority, but are also viewed by their colleagues with a kind of suspicion, as if their loyalty toward the one thing that matters – the pure "advancement" of knowledge – is in doubt.[16]

The debate between Latour and Stengers will pick up again on the occasion of a new proposition, coming from the former, that of the Parliament of Things, which was initially intended to suggest a precise alternative to the old separation of science and politics, i.e., the former epistemology, and thus open the way for Gaia. So far, the appeal has been to values that would complement the facts while remaining separate from them. From now on, it will be a matter of moving from "science-without-politics" to "science-with-politics":[17]

By the appeal to values, we mean first of all that other propositions have not been taken into account, other entities have not been consulted – propositions and entities that seemed to have a right to be heard. Every time the debate over values appears, the number of parties involved, the range of stakeholders in the discussion, is always extended. With the expression "But still, there's an ethical problem here!" we express our indignation, as we affirm that powerful parties have neglected to take into consideration certain associations of humans and nonhumans . . .[18]

The Parliament of Things was only mentioned at the conclusion of *We Have Never Been Modern*, but it is at the center of *Politics of Nature* (1999). To get a good fix on this Parliament of Things, it has to be understood that Latour could just as easily have called it a Parliament of Causes. He takes up the identification between *thing* and *cause*, as Spinoza had done. In French, "chose" (thing) and "cause" have clearly the same etymology, *causa* in Latin. The consequence is made perfectly clear in the Glossary included as an appendix to the *Politics of Nature*: "THING: We are using the term in the etymological sense that always refers to a matter at the heart of an assembly in which a discussion takes place requiring a judgment reached in common – in contrast to 'object.' The etymology of the word thus contains the index of the collective (*res, ding, chose*) that we are trying to assemble here."[19]

This Parliament should allow for the rehabilitation of those we learnt to hate under the old Constitution: "All [epistemologists and political scientists] had in common a hatred of intermediaries and a desire for an immediate world, emptied of its mediators."[20] The function of the Parliament of Things is to put an end to short circuits; to *slow down* to represent better. Latour specifies that the Parliament is not built from scratch, but reunifies what we wanted to separate: sciences and techniques on the one hand, politics on the other. So we have to stop being clandestine! The two have to be "butted together." No voice should go unheard. Those who enter the Parliament of Things are not naked random citizens but representatives "attached" to what they depend on.[21] They are not participating in the creation of an "opinion" but in a confrontation that might have permanently challengeable outcomes. Tolerance is never celebrated here, but all are welcomed into the collective along with the requirements and obligations that go with their attachments.

The first departure from the former Parliament comes with the distribution of the chambers. Instead of a parliamentary chamber on the one hand, and a collection of scientific forums on the other, and a technocratic arm between the two, preparing the negotiations and summing up the facts, we have a single chamber where all the spokespeople gather, whatever the origin of their mandate. The scientific spokespeople are no different to the others, except that they engage in discussions concerning nonhuman constituents, whose capacities and degrees of resistance they can define . . . The parliament looks much more like a laboratory than a House of Representatives, but this laboratory looks even more like a forum or the stock exchange than a temple of truth . . . This transformation brings in a kind of reciprocal right of reply of scientists with politicians and vice versa, completely overturning the former face-off of expert and decision maker . . . At each stage of this changing scale, a certain kind of relation to the future is defined. This is the task of this new political cautiousness: stopping irreversible choices being made on the wrong scale; stopping, in the other direction, experimental decisions being thrown into doubt at the wrong moment.[22]

The Parliament of Things brings science and politics together at its heart; they can no longer allow themselves to "work *separately*."[23] But how can science and politics be mixed? "Without scientific instruments, the political body will never know how many weird entities it has to take into account; without politics, this same political body will never know how to arrange, order, classify this dizzying number of powers to act with which it has to progressively compose a common world."[24] So the Parliament of Things will now have two chambers, not one for science and the other for politics (as under the old Constitution), but one which defines "*how many entities have to be taken into account* (in other words, perplexity and consultation); and *how they can viably live together* (in other words, hierarchies and institutions)."[25]

When some new being – a virus, for example – appears, it is impossible to pretend that it belongs to a "context," or to "the environment" – while it circulates from mouth to mouth, or slips into the most intimate encounters – and expect scientists to take care of it while remaining outside of the arenas of traditional political discussion. It is there, whatever the cost! Scientific controversies get mixed up with political ones, and it is impossible to rely on "the Science" to make everyone come to agreement. We saw the absurd face of the old Constitution when, at the beginning of the COVID-19 epidemic, politicians claimed to rely on "the Science" as they made decisions. They were quickly disillusioned. It was impossible to carve out a domain of solid facts (the scientific domain) in opposition to a domain of negotiable political decisions. The mixture of humans and nonhumans takes no notice of this division. Now the question of ecology can be put in new ways. Latour began to take this step in a 1995 article:

> what would a human be without elephants, plants, lions, cereals, oceans, ozone, or plankton? A human alone, much more alone even than Robinson Crusoe on his island. Less than a human. Certainly not a human . . . the regime [*cité*] of ecology [here Latour is referring to Luc Boltanski and Laurent Thévenot's "cities"] simply says that we do not know what makes the common humanity of human beings and that, yes . . . without the water table of the Beauce, they would not be human.[26]

Stengers, in *Cosmopolitics* (written before the publication of *Politics of Nature*) read this article as a turning point. She opens her "Calculemus" chapter with a long citation, stressing that the grandeur of this ecological regime is the "we don't know" attitude. And she adds, "In Latour's ecological regime, the scientist is 'petty' who is unconcerned about leaving her laboratory to assume the role of an expert, not caring about what the purification required by her practice has led her to ignore."[27] This, to her, means that the Parliament of Things introduced in *We Have Never Been Modern* can become "good to think with." She will later write that its transformation by Latour should be understood and celebrated as a successful "speculative gesture."[28]

If network theory has clearly not convinced Stengers because she privileges "our" networks, she will thus embrace the Parliament of Things, but modifying it in a more speculative direction, playing the ecological regime's "we don't know" off against the attachments needed for, and capable of giving value to, the *We Have Never Been Modern* Parliament.

> What becomes of the Parliament of Things if the grandeur it celebrates and which justifies it is not the political grandeur of constructing ever longer networks, but the ecological and

cosmopolitical grandeur that subjects the ever-renewed relationships between ends and means that networks invent to the test of the "dreams, fears, doubts and hopes" of others, those . . . who know that if the Drome stops meandering, if we banish the bears, if we ban the veil in our schools . . . something of the "common humanity of man" risks being destroyed?[29]

Latour has also stressed the redistribution of means and ends that the ecological question demands: "Ecological crises, as we have interpreted them, present themselves as *generalized revolts of the means*: no entity – whale, river, climate, earthworm, tree, calf, cow, pig, brood – agrees any longer to be treated 'simply as a means' but insists on being treated 'always also as an end.'"[30]

For Stengers, it is never a question of *absorbing* other cultures. So she will add a twist to Latour's idea: the Parliament of Things becomes the cosmopolitical Parliament. Stengers' cosmopolitics has nothing to do with Kant's (she even said that she was unaware of Kant having used the idea when she came up with it: "the term 'cosmopolitics' just came to [her] without warning."[31]). "'Cosmos' means the inappropriable, not in the sense of transcendence but in the sense of what is unknown to politics, conferring on it . . . its capacity to communicate with other modes of consultation, thanks to which other peoples are gathered. This also means separating politics from the claim to constitute a neutral model."[32] She gives a further elaboration here:

> Be careful: none of us has the right to represent "humankind" . . . this formula . . . has a lot in common with the kind of agnosticism for which Bruno Latour has made a requirement for thinking to escape the Great Divide: the "Moderns" are different from the others exactly in the way that they are prone to oppose "humankind" from all the rest . . . the first question for an ecopolitical method described in *Politics of Nature* speaks to this, at the moment of perplexity coming just ahead of consultation: "How many of us are there?", how many different voices have to be brought together for this question? How many different spokespeople because they are coming from a different mode of attachment to the one in question? The question has to be put in the most uncertain, complicated, way, because what allows for simplification can silence ways of knowing that could have, if not should have, counted. A "cosmo" political uncertainty.[33]

An uncertainty that adds to the ecological imbroglios.

If politics defines the way in which the Moderns resolve their differences, there are, as Stengers points out, traditions that could teach us other ways of going about it, African-style palaver, for example:

> In Brussels, I was impressed by the transformations that their constraints produced. By definition, those taking part in palavers are elders; each knows something about the ordering of the world that has to be continued, created, discovered, or reinvented around the case in hand. But the contribution of one person must never take the form of disqualifying what another says. That's one constraint: each recognizes all the others as both legitimate and insufficient – there is a palaver because none of their present knowledge is enough to fabricate the meaning of the situation. And at that point there can be convergences. There is no call for consensus among the participants since everyone, as divergent, is interesting. But, little by little, words that belong to no one in particular start to characterize the situation in a relevant and active fashion. It is this type of experience that interests me.[34]

Choosing the term "cosmopolitics" is Stengers' way of alerting us to the fact that our usual resources (linked to our invention of politics) do not turn us into "angels" capable of spontaneously inventing a "utopia that is valid for every inhabitant on Earth. They [our own resources] do not give us the ability to meet and recognize those who should be the co-authors of such a utopia."[35] It is more a matter of stopping the destruction of those on short networks who know that if the long network carries the day, they will see their chances of survival undermined.

Latour had linked the question of the universal to the length of networks. In *Irreductions*, he wrote:

> When someone talks to me about a "universal", I always ask what size it is, who developed it, and who is projecting it, and onto which wall. I also ask how many people maintain it and how much their annual salaries cost. I know that this is in bad taste [. . .]. And yet, the king is naked and seems to be clothed only because we believe in the universal.[36]

And Stengers, for her part, wrote: "The prefix 'cosmo-' . . . should not be confused with the universal. The universal is a question at the heart of the tradition that has invented it as a requirement and also as a way of disqualifying those who do not refer to it."[37] So this prefix "integrates, problematically, the question of an ecology of practices that would bring together our regimes, where politics was invented, and those other places where the question of closure and transmission has invented other solutions for itself." So it is a matter of thinking of the "coexistence of disparate technical practices . . . characterized by different logical constraints and different syntaxes."[38] "We, who are not angels but think in political terms, must therefore create obstacles that prevent us from rushing toward others while requiring that they resemble what we might become, obstacles that prepare us to wonder about their conditions, the conditions they might establish for eventual exchange."[39] Latour seems to meet her on this point: "universality isn't behind us, to be preserved or destroyed, but ahead of us like backbreaking labor in a vineyard that there aren't enough laborers to do."[40] But Stengers would still wonder about the dreams that inhabit those laborers.

How can one be "capable of escaping the modern passion for disqualifying any practice that fails to accept the existence of a unique world"?[41] She writes, "Sometimes the heterogeneous cannot, even retroactively, be brought back to the homogeneous."[42]

This is why she proposes a new idea: diplomats. Here she pays homage to Leibniz, who "was able to create a conceptual system capable of transforming him into a philosopher-diplomat, never contradicting his multiple correspondents but creating a translation of what they forcefully claimed, which would – if they accepted it – make a possibility for peace exist where a logic of war prevailed."[43] *Calculemus!* She says, dare to "move to the limit [like mathematicians], taking the 'calculation' to its extreme point, inventing its capability to integrate and save what seems to be in contradiction."[44] Here we might be at the heart of the question that counts the most in all of Stengers' *oeuvre*: how to situate propositions that seem to be opposed "such that their coexistence may be not a contradiction to be resolved but a fact to be celebrated."[45] This means complicating the effort to succeed in constructing a *we*, and Latour will pick up on this *Calculemus*:

> Let us not forget the fairy Carabosse! On the pile of gifts offered by her sisters, she put down a little casket marked *Calculemus*! But she

did not specify *who* was supposed to calculate. It was thought that the best of all possible worlds was calculable, provided that the labor of politics could be short-circuited. This was enough to spoil all the other virtues, given how much heroism would have been needed to resist the attractions of that facile approach. Now, neither God nor men nor nature forms at the outset the sovereign capable of carrying out this calculation. The requisite "we" has to be produced out of whole cloth. No fairy has told us how. It is up to us to find out.[46]

The idea of the diplomat is applied in the opposite way from the movement of the dialectic: "Contradictions can sometimes transform into contrasts that it is possible to articulate," she goes on to write. They are "events" that "metamorphose the situation."[47]

The question of diplomats here joins that of ecology (after the transformation of the Parliament of Things into a cosmopolitical parliament). To characterize diplomacy as an

> art, which produces artifices, is a good translation of the fact that those who eventually, possibly, succeed in agreeing were not "made to agree." The agreement doesn't arise from misunderstandings having been overcome or from an equilibrium being returned to, through the resolving of a temporary conflict. If the agreement can indeed "hold," it is not because the world has become "simpler," rid of what perturbed it. If it happens, this is because a world that has become more complicated has been created, sufficiently endowed with enough extra dimensions for what was contradictory to have a chance of coexisting.[48]

"What is at stake in the practice of the diplomat are the effective conditions of an 'encounter,' not the recognition of submission."[49]

Diplomacy takes place among *powers*, so it has nothing to do with tolerance and everything to do with a "translation" procedure. Stengers spells this out in order to avoid any misunderstanding on this diplomacy question: "When someone belligerent undertakes a predatory war, for example, which is to say by defining the adversary as its prey, there is no place for diplomacy."[50]

When the inventors of classical medicines (which have been through clinical testing) attack homeopaths, they can rely on "experts," who have all-purpose ideas, who are not troubled by doubt or fear, and who will get busy sorting out of the "real medicines" and the homeopathic "placebos." The experts typically try to push the homeopaths – those who agree to submit to such conditions – to lock themselves into answers that disqualify them. And the placebo effect will be used here as a weapon of destruction, even though we know nothing about it.[51] Diplomacy is impossible. War is guaranteed. Ironic tolerance is the best homeopaths may obtain. The experts are not troubled by the fact that the homeopathic experience continues to exist, against all odds, supported by both physicians and patients. How can this insistence be taken into account?

Homeopaths, on the other hand, have some chance in the Parliament of Things. They will not be required to submit to the all-purpose drug evaluation criteria as defined by the states in alliance with the pharmaceutical industry, but they will be required to advance proposals about the criteria they themselves would accept as relevant: "You who are presenting yourselves at the door of the collective, what are your propositions? To what trials must we submit ourselves to make ourselves capable of understanding you and getting you to speak?" Then: "Who can best judge the quality of your propositions? Who can best represent the originality of your offer? By what reliable witnesses can you have yourselves represented most faithfully?"[52]

However, it could be that the demand to be represented by "reliable witnesses" subjects them to a temptation for "proof" that, according to Stengers, paradoxically defines the modern charlatan. It could be that the "felicity conditions" as Latour puts it in his *Enquiry into Modes of Existence*, that these questions presuppose are not suitable for some, especially for those whose obligations attach them to a "short network." Stengers' diplomats are there to shake up the "we don't know," to counteract the possibility that the "propositions" or "offers" with which candidates must present themselves do not imply their capacity to integrate into a "long network."

Will we agree not to judge, not to disqualify those who are not interested in such matters, and to accept a world that is a little more complicated and not one where everything that does not fit into the categories made to bring people into agreement can or must be eliminated? Diplomacy depends on an "interstitial possibility, never guaranteed, between what obliges, which cannot be negotiated but only betrayed, and the way in which these obligations are formulated. [. . .] If it is accepted by the different parties, it will be for distinct reasons, not for a common reason that would have prevailed beyond the heterogeneous."[53]

Stengers will combine her diplomat idea with another one, that of "inquirers," which she borrows from John Dewey, and which aims at multiplying inherently political avenues. Dewey proposed that social scientists adopt an "experimental logic" "to learn how a concrete situation is susceptible of changing."[54] Inquirers are first of all "learners"

> who create knowledges that do not designate an object or a fact but a learning trajectory nourished by their own attempts at modification. What is more, the inquirer will be an experimenter in her own manner because the success of the inquiry implies the creation of a successful relationship between the inquirer and the inquired . . . I cannot imagine the Deweyan inquirer dancing on her terrain [Stengers, as we have seen, is referring here to the joy that can rightly seize a laboratory researcher when a daring experiment succeeds] but I can imagine her uneasy joy when those with whom she has worked seize hold of some of the words and knowledge that she has learned with them and make them their own . . . this apprenticeship is not an end in itself, it is correlated with the political event constituted by the emergence of the group that has become capable of intervening, of arguing and of struggling.[55]

Are sociologists still capable of this?[56]

As we have seen, Stengers described Bruno Latour's ecological "city" as speculative. The same applies to her cosmopolitical Parliament. "Speculative" should not be taken in the usual abstract sense, but as that which is less likely to leave us locked in the "probable," to resist it, "to think of something beyond the blind, smothering facts,"[57] to activate "possibilities," which relates to something creative:

> Speculation, in the sense that I understand it, always begins with the insistence of a possibility that makes us feel that things did not need to be conceived as they are, and it tries to nurture this feeling, to explore what it opens up to, what it demands. Its vocation is not to propose but to nourish those who are committed to making things different.[58]

Speculative thinking is opposed to critical thinking, which has the defect of always wanting to show itself to be more intelligent than the actors themselves by bringing to light the forces that act on them without their realizing it. Critical thinking eliminates, claims to know what should be selected from experience and what should be despised. It is thus an avatar of the Great Divide. In opposition to speculative thinking, critical thinking has only contempt for "common sense."[59] Elsewhere, she writes, "speculation deals precisely with the possibility of 'leaps of the imagination,' without which its statements would not make sense."[60] Speculative thinking does not subtract anything from experience, it does not eliminate "any of the dimensions that require, even by a murmur, to be taken into account."[61] Once you begin by eliminating the murmurs, you become deaf to the shouts!

Latour introduces a pragmatist position between "good sense" and "common sense." These terms are

> set in opposition, in order to replace critical discourse and the operation of unmasking; good sense represents the past of the collective, while common sense (the sense of what is held in common, or the search for what may be common) represents its future. Whereas it may be permissible to force good sense somewhat with venturesome arguments, it is always necessary to verify that one is finally meeting common sense again.[62]

And Stengers specifies that:

> Those who criticize speculations say they are empty, utopian, abstract, but they forget to distinguish between speculation "for" or "against" the world. A speculation that occurs "against the world" dreams of ridding it of all obstacles (everyone getting along, being disinterested, etc.). It is recognizable by the denouncers who endorse it and more or less explicitly call for the shaping of some "New Mankind" that will fulfill the dream. But a speculation is produced "for the world" when, rather than purification, it adds, taking the risk of introducing possibilities, additional dimensions.[63]

Speculation is then a way of resisting the "we have to, we have no choice," the refrain that "signals that what had, to that point, been defined as intolerable, quasi-unthinkable, is in the process of creeping into habits." [64]

Stengers relates that she has often been asked whether the Latourian proposal for a Parliament of Things is reformist or revolutionary. There is no answer to this question, she has replied. The point of this Parliament is to bring about a "deformation" of the present "under the effect of a future whose demands are without limits."[65] It functions "without the 'judgment of God' that designates a difference in level delimiting the external and the internal, or a priori disqualifying a particular interest as 'corporatist.' And to the degree that it mines the stable soil of a line of evidence, and provokes problems wherever solutions are reigning, it constitutes a 'concept,' in Deleuze and Guattari's sense of the term."[66]

The Parliament of Things thus calls for scientists adventuring outside their laboratories or the limited scope of their confrontation with their colleagues or their funding bodies, which the theory of long networks strengthens. It puts scientists plunged into a new ecology under *obligations*. They will no longer be able to appeal to the clichés of "progress, freedom of research, objectivity" to which they have recourse when in their pedagogical relationship with the public. What they represent (in the strong sense of the word "represent") is now confronted with much more awkward challenges. Stengers illustrates this idea with the story of the three little pigs and the big bad wolf (she has chosen, for fun, to do a bit of fieldwork, as Latour advocates): "before listening to the experts discussing bricks and cement," one has to problematize what this solution takes for granted:

> Would it not be possible to invent other relationships with the wolf? On what does the definition of the wolf as a menace depend – that is, the definition of the problem as a "protection problem"? In the Parliament of Things, the first priority would be to research – indeed, to elicit – representatives who can point out the possible distinction between the destructive wolf and other possible wolves.[67]

From this point of view, Latour's last proposition is able to provide at least some comfort for Stengers' explicit project, which is to invent "a different way of doing politics, one that integrates what the city separated: human affairs (*praxis*) and the management-production of things (*technê*)."[68] But, as we have seen, she renamed the Parliament of Things as the cosmopolitical Parliament:

> No unifying body of knowledge will ever demonstrate that the neutrino of physics can coexist with the multiple worlds mobilized by ethnopsychiatry. Nonetheless, such coexistence has a meaning, and it has nothing to do with tolerance or disenchanted skepticism. Such beings can be collectively *affirmed* in a "cosmopolitical" space where the hopes and doubts and fears and dreams they engender collide and cause them to exist.[69]

Exit networks defined by their length! It is always a question here of taking care that westerners do not put themselves in the position of being the ones who decide what the tests are and judge all the others. All aspects of the Great Divide must be questioned by the cosmopolitical approach. The cosmopolitical test forbids reducing to belief what are, for others, existential conditions. It thus describes a transformation of the Latourian proposal:

> For me, it was a question of injecting heterogeneity and an additional risk. We in the West invented politics in its public dimension, in the sense of bringing citizens together. "Cosmopolitics" means that we must be wary of the political goodwill that characterizes us – *let's explain, let's discuss*: with all the goodwill in the world, we risk prolonging the destruction of traditions that are not configured for this kind of public test. In Whitehead, one strangely finds, in the course of a sentence, a "cry" which he attributes to Cromwell: "My brethren, by the bowels of Christ I beseech you, bethink that you may be mistaken."[70] Whitehead must have been struck by this cry and so was I. True, you are full of goodwill, but think that perhaps, by simply imposing the test of public discussion on traditions that do not think of humans as citizens first, you can crush them with kindness.[71]

The cosmopolitical proposal did not go unanswered by Latour. He discussed it in a contribution to a Cerisy colloquium held in 2004.[72] After reminding his readers that he is "a great admirer of Beck's sociology," he cruelly remarks, "The limitation of Beck's approach is that his 'cosmopolitics' entails no cosmos and hence no politics either." And he elaborates: "it's for me crucial to imagine another role for social science than that of a distant observer watching disinterestedly." He refers back to the Valladolid controversy, which saw Europeans asking whether Indians had a soul. At the same time, the Indians wondered if the Europeans had a body. "There are more ways to be other, and vastly more others, than the most tolerant soul alive can conceive." For Latour, Beck's proposal is nothing other than "a gentler case of European philosophical internationalism" where "enemies agree on baseline principles." He adds, "Beck and Las Casas are good guys, but good intentions do not resolve or prevent strife." And he finishes by opposing Stengers' *cosmopolitics* to Beck's *cosmopolitanism*:

> The presence of *cosmos* in cosmopolitics resists the tendency of politics to mean the give-and-take in an exclusive human club. The presence of *politics* in cosmopolitics resists the tendency of *cosmos* to mean a finite list of entities that must be taken into account . . . *cosmopolitics*, in Stengers' definition, is a cure for what she calls "the malady of tolerance." . . . William James's synonym for *cosmos* was "pluriverse," a wonderful coinage that makes its multiplicity clear.[73]

In a book published in 2010, *Cogitamus*, he comes back to the word "cosmos" in its ethnographic meaning: "The arrangement [*agencement*] of all beings that a particular culture binds together in practical forms of life."[74] And, he clarifies, "When anthropologists say 'all beings,' one must have a broad mind and one's heart in the right place: these are gods, spirits, stars, as well as plants, animals, kinships, tools, or rituals." The meeting with Nathan continues to bear fruit. We are no longer just trying to distinguish between science and politics but "among compositions of *worlds*."

This leads him to explicitly claim the notion of cosmopolitics as more speculative than the Parliament of Things: "I borrow from the philosopher Isabelle Stengers the somewhat unusual term of *cosmopolitics* . . . It is about what happens to worlds, and, for example, to viruses, neurons, particles and atoms. Growing disputes about science and technology force us to take this new cosmopolitics seriously . . . the arrangement of the cosmos now depends on public life."[75]

In the cosmopolitical Parliament, where everything must be slowed down, there are not only politicians, who speak for humans, and scientists, who speak for nonhumans, but "also a great many actors, activists, practitioners, consumers, enlightened amateurs, and specialists of all kinds, whose unforeseen skills are entitled to have a say."[76] Networks no longer depend on their length alone. Those who study them are now equipped to differentiate and cross-reference them much more astutely; they must be able to resist the western pretension to judge. A pretension which has become a business-as-usual affair for producers of networks, and they have more than one trick up their sleeves to constantly reinstall themselves in this comfortable position!

7
Identifying Modes of Existence, Thinking with Whitehead

In 2002, Latour took on the law as an object of study, and published an ethnographic inquiry on the Conseil d'État, a French Supreme Court (court of appeals against rulings by administrative tribunals).[1] The counsellors there work in a way which seems sometimes akin to that of the scientists in *Laboratory Life*. But it is mainly his ethnographic method in both cases which creates similarities with scientific work. The counsellors have to be studied as they are "going about their business," and it is very important to discover the mysterious thing that "makes them do what they do" [*les "fait faire"*], what moves them, that is, as Stengers would say, what *requires* and *obliges* them. Like the germs that Pasteur made manifest because he created the right apparatus (that the germs "had him make" – this apparatus had to be suitable for them to respond to it with a "present!" – showing themselves to be actants in their own right), something "makes" the counsellors "do what they do." In every instance, the specific actants have to be identified. At the same time, Latour abandons the distinction between long and short networks in favor of another question: what is circulating *specifically* in each network?

Latour is not doing a sociological study of these counsellors (there is no reason for him to be interested in the sociology of professions), but he is keen to characterize the *objects* that are moving among them in a specific manner. What is it the counsellors "make *pass* through their interactions"?[2] His point is to characterize the "objects of value" and their felicity conditions (and reciprocally, their infelicity conditions) which are circulating through the practices of the Conseil; what are called "means" in juridical language. In order to do this, the counsellors have to tirelessly reconsider prior decisions taken on comparable subjects:

> law is indeed a mode of exploration of being-as-other, a particular mode of existence . . . it has its own ontology . . . it engenders the human without being made by them . . . it does belong to the prestigious category of "factishes." It does what no other regime of enunciation does: it keeps track of all disengagements, to tirelessly reconnect statements to their enunciators, via the perilous route of signatures, archives, texts, and files.[3]

Latour was to add to his first inquiry (on religious speech in his 1975 thesis, complemented by his little book *Jubiler ou les tourments de la parole religieuse*, published for the first time in 2002), another on "psychism" at the Georges Devereux Center (where he identified "transfearances" that would later become "metamorphic beings") and a study of the law (and the very particular way that the counsellors at the Conseil d'État work) to show that these beings (that *make demands* or *"font faire"*) emerging from scientific practices typical of laboratories cohabit with many others that are quite different but equally respectable.

From his very first study on religion, what is striking is the specificity of religious speech, so that it would be stupid to try to see it as the same as laboratory work: "[T]he more a layer of texts is interpreted, transformed, taken up anew, stitched back together, replayed, and rewoven, each time in a different way, the more likely it is to manifest the truth it contains – on condition . . . that one knows how to distinguish it from a different mode of truth."[4] He elaborates later:

> In natural or social science, the researcher has a duty to add his stone to the vast edifice of knowledge, to discover, to innovate, to produce new information; but in matters of religion, his duty is faithfulness: he doesn't have to invent but to renew; he doesn't have to discover but to recover; he doesn't have to innovate but to revive the never-ending refrain afresh.[5]

Identifying Modes of Existence, Thinking with Whitehead 99

If there is one thing in common among religious speech, laboratory work, and the fabrication of the law, it is only the "long struggle against the eradication of mediations,"[6] an eradication that we have seen at work in epistemology, and which he will call the "double click": "Double-click communication, this immediate and costless access, this conveyance that appears to demand no transformation, has itself become, for our contemporaries, the model of all possible communication, the ideal, the metric standard of all movement, the judge of faithfulness, the guarantee of all truth."[7] This battle against the eradication of mediations has to be carried out everywhere so that each "mode of existence" can appear in its singularity.

Latour will then get on with the task of characterizing and classifying all the beings dear to the hearts of the Moderns, although they may not be clearly aware of them, *according to their particular modes of existence* – each mode, especially the scientific, but also the religious, tends to "the hegemonic, and to being unaware of the others" – all these beings encountered during his multiple inquiries and which have come to populate the cosmopolitical landscape.[8] This will result in the publication in 2012 of *Inquiry into Modes of Existence*, a book it took him twelve years to write. Here he shows how the mentality of "critique" (or, as he also puts it, "the ordinary common sense of the social sciences") is mistaken in believing that it has to find something hidden. There is nothing to be found, neither behind fiction, nor the law, nor science or religion. One only has to return its *dignity* to each individual mode of existence, taking good care not to mix them up. He even borrows Foucault's expression "regimes of truth," which has multiple similarities with his felicity conditions. In the conclusion to the book, he sums up how he moved from what are called "domains" to "modes of existence":

> we choose, among the Moderns, the DOMAINS to which they seem to hold the most; we shift attention from these domains to NETWORKS; then we look at the way the networks expand, in order to detect the distinct tonalities that we gradually extract by comparing each network with the other modes of extension, two at a time; finally, and this is the hardest part, we try to entrust ourselves exclusively to the often fragile guidance of these discontinuous trajectories.[9]

What enabled this book to take the form it did was the encounter with the forgotten work of the philosopher Étienne Souriau and especially his *The Different Modes of Existence*, first published in 1943. This will give us the only text jointly written by Stengers and Latour (apart from some open editorials), "The Sphinx of the Work," the preface to the new edition of two Souriau essays.[10]

But before this Stengers produced a huge book which we shall have to spend some time on. It is the outcome of her early encounter with a philosopher, little known in the francophone world, with whom we have crossed paths from the start of this book, indispensable as he is to understanding our two authors: Alfred North Whitehead (1861–1947). Her book is not *on* Whitehead, it is written *with* Whitehead, a mathematician (this is not without importance) who turned to philosophy, and with whom she shares an intimacy that makes them almost co-authors of her book. Thus she has no hesitation in extending the texts he left behind: "As long as Whitehead's text remains alive, it will transform its readers into co-authors, accepting the adventure of the imagination to which the text calls us."[11]

On publication, Latour praised the book: "the greatest philosopher of the twentieth century is finally studied in great detail by someone who is one of the most innovative philosophers of science of the present time . . . Stengers illuminates the most obscure passages of Whitehead in a style that is subtle, often witty, always generous."[12]

Latour could have inherited something from Whitehead because, as Stengers writes,

> reading Whitehead, means always, of course, posing the question which accompanies the history of philosophy like its shadow, "What is philosophy?", but here it is not a matter of celebrating a history that makes progress and giving up coincide. For Whitehead, no one has to give up anything, for whatever reason, and nothing that happens to the few has any claim to happen for everyone. For instance, in a quite classical fashion, Whitehead links the question of modern philosophy to the rise in power of scientific disciplines as unchallenged authorities in the field of knowledge, and to the decline of religion as a moral authority, but he will never double up "what happened" with a reason for why it had to happen. . . . For Whitehead, thinking through the modern age means thinking

through the triple problem of the triumph of the sciences, the decline of religion, and philosophy's retreat into "subjectivism"; in other words, the (exhilarating or depressing) task that philosophers have slowly taken on of thinking through the only questions that a scientist is not supposed to be able to ask.[13]

Latour makes use of a Whiteheadian concept – "bifurcation of nature," between "real nature" and "apparent nature" – which we have already encountered, but which we now have to focus a bit more on. First, let's consider the way in which Stengers presents it: "On the one hand, there is an objective nature, ruled by causality. It causes our perceptual experience in particular. On the other hand, there is nature as we perceive it, rich in sounds, colors, and odors, as well as values, emotions, fear, and wrath. It is mere appearance, for which the human mind alone would be responsible."[14] In *Thinking with Whitehead*, Stengers takes this bifurcation as her (and Whitehead's) starting point. In her introduction, she writes:

> It is this question, under the name of the "bifurcation of nature," which constituted the initial stakes for Whitehead's adventure into speculative philosophy. Whitehead did not denounce or criticize, or seek historical explanations for the fact that our thought, our theories, our words, make nature bifurcate into an "objective" nature and the bundle of values, meanings, beauties that we have to learn to attribute to ourselves. From the very start, he defined this bifurcation as "absurd"...[15]

Absurd, as when Prigogine refuses the irreversibility of physicists' time. Again, she says that this is "a real phantasmagoria":

> The first thinkers who suggested distinguishing an "objective" nature, characterized in terms of so-called "primary" qualities ... from the nature we actually experience, rich in smells, colors, meanings, were certainly adventurers. But today the distinction has become a "slogan," disseminated as if it were obvious, a vector of absurdity, a producer of dead-ends that are not only intellectual, but also practical, even political.[16]

This is not something to be denounced, but one should learn to resist it: "Philosophy, for [Whitehead], is not denunciation. Yet it can indeed be said to be resistance."[17]

If, as an abstracting operation, the bifurcation of nature is indispensable to the experiments taking place in laboratories (at the same time as revealing its limits because a successful "take" in this well-defined and singular context that is the laboratory cannot problematize the "diversity of kinds of grasp, modes of abstraction to which nature is susceptible"[18]), it becomes a caricature in the so-called human sciences where it goes by the name, as we have seen, of *critical thought*, revealing a hidden truth behind appearances.

She comes back to this during a Cerisy colloquium:

> That for which the responsibility can be attributed to "nature" will be called "objective," and which has, as such, the entitlement to silence disagreements and conflicts, and banish them to the subjective. So bifurcation is part of the hidden agenda [of the modernity that defines us as those who have to be protected against our own irrationality, kept on the straight and narrow] which has constituted "science" as maintainer of law and order.[19]

For Stengers, the question of bifurcation moved from the philosophical register (Kant elevated it into a doctrine) into the political one. It extended the exceptional rather than exemplary fact of laboratory success to everything popping up in the guise of Science (capitalized in the singular, "advancing like a great monotonous wave,"[20] to be distinguished from the sciences in the plural), and which would only be a matter of choosing the "right method." The example she recalls for us is that of IQ measurement that Stephen J. Gould spoke of as a "mismeasure." In this way, Science "bifurcates everything it touches."[21] "The mode of abstraction invented in the laboratory has the singular value that it presupposes that every attempted take can fail."[22] As for the repetitive "sciences" that do not invent anything – they cannot fail!

The theme of the bifurcation of nature allows us to return to *speech impediments* (or the *middle voice*) with which we began this book. Bifurcation "asks who is responsible for what we know, either the 'objective' agents of nature, or our 'subjective' assessment methods."[23]

Whitehead turns to *vigilance* when faced with the potential ravages of modes of abstraction when they are in denial, or when they revert to appearances, to subjectivity, to what has

Identifying Modes of Existence, Thinking with Whitehead 103

been ignored, or excluded, in order to guarantee their success. A science only "deals with half the evidence provided by human experience."[24] One has to be vigilant in relation to the "predatory power that modernity has conferred on some of its modes of abstraction."[25]

Stengers imagines and calls for specialists who would be capable of linking "actively what they know and what their knowledge, to be produced at all, must leave out."[26] The biologists who created GMOs should recognize "that in the fields things can happen that cannot be observed in the laboratory"[27] and should not count on Progress to sort things out . . .

If Whitehead is so important, it is because his refusal of the bifurcation of nature places an obligation on philosophy "not to explain by elimination, depriving an experience of its own value."[28] According to Stengers, one has to refuse the "Sacred Union" surrounding scientific facts under attack, "whatever they may be." "Whatever they may be": what this expression is targeting is those who refuse to criticize the pseudo-facts produced by many sciences because it might weaken the power of facts to "keep the people silent,"[29] as Latour puts it or, in Stengers' words, "silencing the plurality of the partial and discordant opinions of the irrational populace."[30] One should not give priority to "any particular form of knowledge."[31] The question preoccupying Whitehead is

> the lack of resistance characteristic of the modern epoch to the intolerant rule of abstractions that declare everything that escapes them frivolous, insignificant, or sentimental. We must take seriously the fact that the disasters of the modern world, its ugliness and its brutality, are part of what is called "progress"; what has been done was done in the name of progress."[32]

Thus the bifurcation of nature enables us to get a new perspective on the Great Divide question. In the conclusion to her book, Stengers calls for a different kind of "ethical" scientist:

> physicists capable of celebrating the adventure they inherit in its singularity, without turning the "physical reality" of the electromagnetic waves emitted by the sun into "the" objective version, in opposition to which all the other versions must be defined. Such physicists may, in the trajectory that gave them this capacity, have been the fundamentally anonymous site of experiences that testify

to evil, *qua* liable to be overcome. And they may, of course, by the example they provide, be accused by their peers of "demoralizing" the community. This happened, for instance, to Henri Poincaré and Pierre Duhem. Yet their "ethics" remain indeed those of physicists, as do their dreams, their doubts, their hopes, and their fears. They have "simply" acquired the good habit of dreams that do not turn them into the thinking head of humanity, taking charge of the questions that "men" have allegedly asked themselves forever, and which physics would recently have learned to answer.[33]

But now we have to go back to an episode narrated by Latour himself: "She [Stengers] challenged all my sociosemiotic developments with a vigorous 'I know, I know, but even so . . .' and, making a characteristic rapid circular movement with her right hand, demanded that something be brought to the surface in the analysis, something that would be the world but grasped differently." Latour goes on to describe the scene:

> I am almost certain that it was in 1987 [therefore after having written *La vie de laboratoire* and the English version of *Science in Action*], during a conversation by the swimming pool at Fondation Les Treilles,[34] in Provence, that Stengers shared with me an astonishing quotation from Whitehead . . . about the risk taken by rocks – yes, rocks – in order to keep on existing . . . Stretched out in the sun on an island across from Gothenburg, in Sweden, I could not stop running my fingers over the rough red surface of the rocks as if to find out whether Whitehead could have been right! . . . There exists a completely autonomous mode of existence that is very inadequately encompassed by the notions of nature, material world, exteriority, and object. This mode shares one crucial feature with all the others: the risk taken in order to keep on existing.[35]

And, he adds, "From that starting point, everything quickly fell into place . . . I was able to chart in one fell swoop the regimes that I was going to have to investigate more systematically."[36]

The mode of existence that Latour is talking about here (designated by the preposition [REP] for reproduction) will have a decisive role to play in the *Inquiry into Modes of Existence*. It is, as we shall see, the mode of existence without which Gaia cannot be thought. But before this book was published, Stengers and Latour wrote a text ("The Sphinx of the Work") which we have already mentioned and which could be considered preparatory.

It appears as the preface of the new edition of Souriau's works, *The Different Modes of Existence* and *Of the Mode of Existence of the Work-to-be-Made*.

According to Stengers and Latour, Souriau's aim is to proceed to a "prodigious expansion of empiricism."[37] While the "world [was] reduced to two modes – object and subject," objective reality and knowing subject, he suggested abandoning this rickety epistemology in favor of the *instauration* of multiple and different beings: "the soul as well as the body, the work of art as well as the scientific entity, an electron or a virus."[38] We can understand how Souriau's project can sit coherently with Whitehead's refusal of the bifurcation of nature. And we can better understand the importance of the "How should I put this?" in the introduction to this book.

The word "instauration" resonates with "factishes": it "engage[s] with the question of the work in an entirely opposite way to constructivism,"[39] and it allows us to think through the relation of reality and the work. It allows "transactions with many other types of beings, in science, in religion, in psychology, as well as in art."[40] For a long time, Latour had been looking for the right word; first, he had tried "to seat" [*asseoir*], but *instaurer*, Souriau's idea, brought his search to an end.[41] Each mode of existence allows one to "bite" on a part of the world: "there is more than one dwelling place in the kingdom of experience."[42] Each mode of existence is a key to interpretation.

Instauring demands the determination of the mode of existence that will be the interpretive key that specifies it, its efficacy, its "anaphoric mode" – its "monumentality," says Souriau – which is to say the way it maintains itself via its successive appearances, its continual transformations, and its trajectory of reprises[43] (Latour will later say that one has to "grasp continuity through a series of discontinuities,"[44] of alterations – in order to grasp the being-as-other). Each mode is "an art of existing unto itself,"[45] each has a "particular pattern," as in dressmaking, that cannot be applied to others, each has "a different way of undergoing the trial of anaphor,"[46] a little as if each "form leaves in its wake a different way of 'having a soul,'"[47] where there is always the risk of being lost (Latour also speaks of "felicity and infelicity conditions"[48]). For each mode, we have to ask ourselves what are "the unique factors of reality."[49] We have to learn "to identify, for each type of practice, the rich vocabulary that it has managed to develop to distinguish truth from falsity in its own way."[50]

Each mode corresponds to a test of anaphora that is specific to it. The opposite of anaphor (the repeated movement of prolongation and reprise) is what Latour in his *Inquiry into Modes of Existence* calls Double Click, an Evil Genius idea which inspires lethal interpretive mistakes everywhere. Epistemology is the perfect example of it, in that it consists of forgetting or erasing *mediations*: "Poor Double Click is starving for everything: politics, religion, law – won't he end up starving to death?"[51] Double Click threatens each mode with incomprehension, because it has a "horror of hiatuses" and offers "displacement without translation."[52] The risk of discontinuity and the constant necessity for reprise, what Latour calls "hiatus," are present each in its own way for every mode. Each is characterized by its mediations. Each of the fifteen modes of existence that Latour identifies in his inquiry has its own kind of hiatus, its own trajectory and form of veridiction (there is no Truth here, but rather specifically variable veridictions, as James also suggested). And each mode is fragile, under threat, and the anaphor can always fail. It is a "veritable battlefield,"[53] and the risk is one of "dissolution in nothingness"[54]; "the scales on which these modes are successively measured is their relationship with instauration: each one represents a particular degree of risk, a risk in which the success or failure of anaphoric experience is demonstrated with more and more clarity."[55]

Souriau took fictional beings as one example; they need our "solicitude."[56] With the mode (FIC), Latour picks up on this directly. This mode allows us to understand, very directly, what a mode of existence is and the need not to confuse it with others, especially those characterizing scientific practices ([REF] for "reference"). These ghosts and chimeras can be endowed with more reality than the "Duponds or Dufours with whom we are summoned to coexist."[57] We then have to define how these special beings – from Don Juan to Tintin passing by way of Mme Bovary – "exist and . . . cause us to exist."[58] Without confusing them: "Like a mother who loves all her children with the same exclusive and all-embracing love, the investigator has to reconstruct each mode's exclusive manner of demanding its truth."[59]

Latour and Stengers have suggested that researchers no longer present their work in reference to Reason and Progress, nor to the opposition between the rational and the irrational. This is the outcome of all those years Latour spent studying how objectivity is fabricated – and which has unfortunately been confused with relativism. Latour wrote that it was a matter of scientists shifting "from Certainty to Trust," both engaging with "entirely different philosophies, or rather metaphysics, or better still, ontologies."[60]

But one must take care not to understand this overall architecture as linking up all the modes "in an harmonious totality." This is not a repetition of the Hegelian dream of totality, but there is nonetheless an architectonic, meaning the possibility of regrouping the different modes in larger categories. Everything is organized, but everything remains open. Latour's *Inquiry into Modes of Existence* is indeed conceived along these non-totalizing lines, as a "work-to-be-done": the reader is asked to extend it through the particular device of an internet site enabling collaborative research, a crowdsourcing of readers invited to become researchers.

With his modes of existence, Latour is closest to Deleuze and Guattari. In her *Thinking with Whitehead*, Stengers cites a very telling extract from *What is Philosophy?*:

> There is not the slightest reason for thinking that modes of existence need transcendent values by which they could be compared, selected, and judged relative to one another. On the contrary, there are only immanent criteria. A possibility of life is evaluated through itself in the movements it traces and the intensities it creates on a plane of immanence: what does not trace or create is rejected.[61]

From this point on, everything is in place to repopulate or replace the world that the Moderns have until now reduced to a dispiriting and sterile face-off – between nature and society, subject and object – with the world we really live in. As always with Latour, everything begins with the necessity for a good description, which can only come out of well-run inquiries. Latourian philosophy is radically empiricist.

We are not alone in the world. Stengers says that Latour, with his *Inquiry into Modes of Existence*, turned himself into a diplomat. He laid a bet, "the possibility that what we call our civilization may have a future."[62] Yet one does have to take into account that nothing is further from being a diplomatic figure than the philosopher Souriau, haunted by "the weighty responsibility that all the other beings of the world impose on 'accomplished individuals, at the height of their powers.'"[63] Who else could this be "if not the (white) man taking on his shoulders, yet again, the noble burden of the role of guardian of humanity's progress?"[64] In a quite different way, Latour imagined an agora

> would gather protagonists concerned by the possibility of formulating ways of speaking well about forms of knowledge locked in rivalry today, each trying to disqualify the other . . . Scientists of diverse provenance will be there, as well as other practitioners, legal experts, theologians, and doctors, to name a few of those who belong to institutions that tend to hold public self-presentation at arm's length. And since diplomacy is an apparatus, it is all about seeing how it can function. How is the assembly to be composed? How will roles be distributed? . . . The wager of Latour, diplomat, is that each practitioner is accustomed to presenting her practice to people deemed incompetent and incapable of understanding and is primarily concerned with maintaining an apparatus of "territorial defense" . . . Latour's agora, moreover, requires the presence of the public in a mode analogous to that of the Athenian citizens, whose civic duty was to attend the spectacle of passions, staged through tragedies, and brood with the chorus commenting on the events . . . It is vigilant . . . The speculative character of Latour's agora underscores the absence of such apparatuses in the modern world, apparatuses nourishing a political culture capable of distinguishing between what we call democracy and the art of leading a herd.[65]

If Souriau's "beings" have been "enlisted" on Latour's diplomatic mission, Stengers will insist, in a piece she wrote in 2015, on an important difference:

> The very meaning of the assertion "there are beings," the key to Souriau's philosophy, has mutated. These beings are no longer required to "speak well" of the experience of a work demanding its accomplishment, but by a more generalized misspeaking. Where Souriau was thinking of a plurality in harmony, Latour's thinking is forced by a desperate and rabid cacophony of misplaced pretenses and sneering mystifications.[66]

What she stresses is:

peoples we have judged to be backward are entitled to diagnose us as being appallingly reckless, with regard to the relationship we have with the beings that make us think, imagine, and feel . . . The beings we have come into contact with have mutated into rapacious powers, blindly claiming a right to exist in a way that turns those who serve them into predators.[67]

Latour, then, knew how to "instaure with appropriate caution," which is to say, to use his own words, to instaure

> Beings whose continuity, prolongation, extension would come at the cost of a certain number of uncertainties, discontinuities, anxieties, so that we never lose sight of the fact that their instauration could fail if the artist didn't manage to grasp them according to their own interpretive key, according to the specific riddle that they pose to those on whom they weigh; beings that are standing there, uneasy, at the crossroads.[68]

One has to get away from what Souriau called the "monumentality" of beings. Latour "deliberately accentuates" the vulnerability of the paths of instauration. Instead of "conquering trajectories," he has "fragile instauration pathways, always to be taken up again. He does not speak of 'beings in themselves,' whose demands we (always we . . .) will finally have understood. Latour's idea is one of the art of relationships, and the uneasiness of the beings speaks of a situation that has to be accepted in this respect – an apprenticeship situation."[69] This was not the case with Souriau's beings, which

> ask to come to their truth, and call to "man" whose duty is to respond – in this we have a grandiose form of humanism, which violates the limits assigned to human finitude. But it is for the same reason that it is a question of betraying him, or perhaps of hearing him as asking us the question of what, at the time of the Renaissance, was called "humanism" before it was buried and forgotten by the figure of the self-sufficient human, conquering superstition, fetishism, the "animist" beliefs of those who did not think they were alone in the world.[70]

Deleuze drew the same lesson from Souriau:

> So it is not a matter of calling and duty – Deleuze never spoke of the "work-to-be-made" as the fulfillment of some cosmic achievement,

he says he just created a few concepts. And this creation does not bear witness for the human race . . . Deleuze never thought of "man," nor the Work, and if he was interested in life, it is in a larval, metamorphic way . . . He was not opposed to Souriau, but as he drew from him he used the language of betrayal.[71]

So Souriau has to be betrayed![72] This comes as a thunderclap for all those who had rediscovered Souriau a little too quickly and were using his work. One has to give up on "the architectonics of the monument that Souriau erected for the future." This is the meaning of Latour's diplomatic mission that goes by way of "a culture of care and carefulness."[73]

But Stengers still has a lingering doubt:

Today, the agora as Bruno Latour imagines it might appear to belong to a dead past. Diplomats could have intervened at the time when the science wars were waged, when the public still paid attention to specialists. But is not such a public on the verge of extinction? Panic now overcomes the herd previously held in check by promises of progress and economic growth, as well as the inexorable advance of knowledge that placed the planet at the service of humans.[74]

Catastrophic times; facing Gaia.

8
The Intrusion of Gaia

When Stengers published *Au temps des catastrophes* in 2009, she wanted to call it *L'Intrusion de Gaïa*. It was her editor (myself, as it happens) who talked her out of it... The same thing happened to the title of a lecture given in 2012 at Saint Mary's University in Canada (finally called "Cosmopolitics: Civilizing Modern Practices"[1]). This is why I have given this chapter the title that was missed twice before. As the activists say, *Reclaim it!* We are going to try to go back along the path that led both our authors from their earlier work, particularly on the sciences, to Gaia. Gaia, the ticklish one:

> Gaia is the name of an unprecedented or forgotten form of transcendence: a transcendence deprived of the noble qualities that would allow it to be invoked as an arbiter, guarantor, or resource; a ticklish assemblage of forces that are indifferent to our reasons and our projects... Gaia makes a major unknown, *which is here to stay*, exist at the heart of our lives.[2]

Gaia is the one who with a "'shrug of the shoulders' [is] capable of making us lose our foothold."[3]

It is with Gaia – which he inherits from James Lovelock under the gaze of Whitehead and his well-named "philosophy of organism" – that Latour will resolutely enter politics. Gaia did appear in a more surreptitious way in *The Politics of*

Nature. The *Inquiry into Modes of Existence*, an inquiry that he asks to be constantly reprised, marks a key stage in the huge ethnographic work on the Moderns that began with his science studies. He succeeded in describing the Moderns positively. In the introduction, he specified his aim:

> only then, might we turn back toward "the others" – the former "others"! – to begin the negotiation about which values to institute, to maintain, perhaps to share. If we were to succeed, the Moderns would finally know what has happened to them, what they have inherited, the promises they would be ready to fulfill, the battles they would have to get ready to fight. At the very least, the others would finally know where they stand in this regard. Together, we could perhaps better prepare ourselves to confront the emergence of the global, of the Globe, without denying any aspect of our history. The universal would perhaps be within their grasp at last.[4]

And, he added, "the touchstone that served to distinguish past from present, to sketch out the modernization front that was ready to encompass the planet by offering an identity to those who felt 'modern,' has lost all its efficacy. It is now before Gaia that we are summoned to appear."[5]

There is no break between all of Latour's earlier inquiries and his ecological engagement. If we glance back at his first works that bring out the actions of nonhumans, he writes:

> and besides, we were right! History – I mean the history in which ecology was about to force the association of humans and nonhumans to be taken into account – was about to prove this. Here, at least, no one could take us by surprise; equipment in hand, we were waiting for this new world or, to put it better, we were waiting for it like the servants in the Gospel, our lamps already lit.[6]

Latour's ideas will overturn the field of ecological studies, and be extended by numerous other researchers into the most diverse fields. His *Facing Gaia*, first published in 2015, represents a major step forward. Some people, thinking him to be a relativist, even thought of this as a reversal: now he is defending the IPCC and the scientists working on what he calls the New Climatic Regime?[7] Nothing is further from the truth. *Facing Gaia* would not have been possible without all those years spent on understanding the sciences and the Moderns. Latour points it out himself:

> I have never thought that "facts" were objects of belief, and because, ever since *Laboratory Life: The Social Construction of Scientific Facts* (with Steve Woolgar, 1979), I have described the *institution* that makes it possible to ensure their validity in place of the epistemology that claimed to defend them, I feel better armed today to help researchers protect themselves from the attacks of negationists. It is not I who has changed, but those who, finding themselves suddenly attacked, have understood to what an extent their epistemology was protecting them badly.[8]

Already in 1984 he was writing in *Irreductions*:

> What would happen if we were to assume instead that things left to themselves are lacking nothing? For instance, what about this tree that others call *Wellingtonia*? Its strength and its opinions extend as far as it does itself. It fills its world with gods of bark and demons of sap . . . The tree gives of its forces, and as it does so, it discovers what all the forces it welcomed and interpreted can do. You laugh because I attribute too much cunning to it? . . . Will you deny that it is a force? No, because you are mixed up with trees however far back you look, or extend your gaze. You have allied yourselves with them in endless ways, to the point that you cannot disentangle your bodies, your houses, your memories, your tools, and your myths from their knots, their bark, and their growth rings. . . . You claim that you define the alliance? But that illusion is common to dominators and idealists of all stripes. . . . And, if you are all filled up with trees, how do you know they are not using you to achieve their dark designs?[9]

This is already a eulogy for what Stengers will designate as *involution*, in the sense that the theory of evolution should be complemented by a theory of involution: the branches of the evolutionist tree become crossed and entwined (the best known example of these "species assemblages" is the wasp and orchid that Darwin already celebrated).[10] Here we are already among the interdependencies that will be at the heart of *Down to Earth*.

In one of his latest books, *After Lockdown*, Latour makes a point of the distinction between *the world one lives in and the world one lives from*, permitting a reconsideration of the colonization question. We have to learn how to land, which is to say to situate ourselves in a territory characterized by all that we depend upon, even if it means assessing what unknown others have to do to satisfy what we think we need. And doing ecology means drawing up a list of all that in order to determine the livability conditions of the world. This is a terribly difficult task because we imagine ourselves in one world, whereas we actually live in another, and this means asking all the political questions differently. We have to begin to learn how to describe the world as "we are experiencing it," Earth, that is, and not as a function of abstract laws that concern only the Universe (which again relates to the bifurcation that nature has been made to undergo). So we have switched worlds, as COVID insists on reminding us. Latour illustrates this point by giving his book the form of a philosophical fable: we are like Gregor in Kafka's *Metamorphosis*, who overnight became a cockroach, with nothing in common with his parents, living like shadows in the former world.

We have seen that by bringing to light what the sciences do in practice, Latour has given their rightful place to the actants, or the nonhuman actors, and this begins in his first book on Pasteur. It is this taking of nonhumans into account that distanced him from the sociology of the sciences (Anglo-Saxon STS folk) and enabled him to characterize himself as a "relationist" in opposition to the relativists.[11] Early on, he challenged the separation between, on the one hand, a "society" composed solely of humans and, on the other, a "nature" composed of inert objects. "Things" are always also "causes." This was verified for everyone in a spectacular manner with the arrival among us of a virus which became the lead actor in a matter of months, capable not only of killing humans but also of turning democratic freedoms upside down, destroying entire sectors of activity, forcing humans to stay away from each other, to stop traveling, to not move more than one kilometer from home (yes! in France), to relegate to the past their beliefs in decisive victories over infectious agents. Who could have imagined a nonhuman to have such power? By giving a central place to reproductive beings (REP) – among all the beings-as-others that his *Inquiry into Modes of Existence* celebrates – that is to say, beings whose concern is to *subsist* and whose felicity conditions are "to continue" and "to inherit" (or as a condition of infelicity, "to disappear"), that is, by redefining what *materiality* is, Latour will have equipped himself with ways to "face Gaia."

For her part, Stengers assiduously crafted differences among the sciences – even preferring, in the end, to speak of "knowledges" – by insisting on the *requirements* and the *obligations* of each and every one, and she showed how a success of the type Galileo inaugurated can constitute a poison. This type of success, even if it should be praised precisely because of its exceptionality, could only take place in a laboratory, but it became a model to be imitated, a simple methodology that can be generalized everywhere. It also became a poison for the researchers who speak of their experience in the terms of a ready-made, transforming what deserves the name of "adventure" into an irresistible "Progress," which all humans would have to bless.

In parallel, she never ceased to do justice to a small book by Félix Guattari first published in 1989, *The Three Ecologies*, which allowed her to approach the question of capitalism afresh, not as a system of exploitation (something she nevertheless retains) but as a system of *destruction*, in particular of practices – for example, practices linked to the commons.[12] This destruction, moreover, is carried out in the name of progress, the fight against those who are accused of wanting to "go back to using oil lamps."

Stengers explains:

> If one is looking to reread Marx today, one should, in my opinion, push, as far as it can go, the anti-Hegelian idea, that many Marxists think they have "accepted," but I don't think they have yet drawn all the consequences from it: nothing of what capitalism has destroyed – corporations for example – can be considered to have helpfully removed an obstacle in the way of socialism; everything it has destroyed must be thought of as *having been destroyed*. There is nothing acceptable about the notion of the "cunning of reason." Let's refuse the idea of a beneficial destruction that simplifies life. What is destroyed is really destroyed. When one is thinking, one should feel one is the inheritor of all that has been destroyed; it should be a weighty feeling. This weight should bring the obligation to think, and oblige one to think against the little refrain of progress.[13]

It isn't the cunning of reason we should be talking about here, but rather Whitehead's "trick of evil" that emphasizes the possibility that a new proposition insisting on its importance "at the wrong season"[14] may generate rigid inhibition and be felt as an insufferable contradiction. If there is one thing that neither

Stengers nor Latour inherit, it is the Hegelian dialectic, even in the version resuscitated by Marx. The idea of diplomacy, which became so important for them, is precisely not about "opposing" the discordant multiplicity but "situating" it in such a way that "coexistence may be not a contradiction to be resolved but a fact to be celebrated."[15] Latour writes:

> We no longer expect from the future that it will emancipate us from all our attachments; on the contrary, we expect that it will attach us with tighter bonds to more numerous crowds of *aliens* who have become full-fledged members of the collective that is in the process of being formed . . . We shall indeed have to involve ourselves still more intimately with the existence of a still larger multitude of human and nonhuman beings, whose demands will be still more incommensurable with those of the past, and we shall nevertheless have to become capable of sheltering them in a common dwelling.[16]

The "season" when the Gaia question arose with urgency was certainly wrong. Those who should have felt what it means and what it imposes were facing other problems. In particular, university people were, and still are, trying to survive and adapt to the "modernization" of their institutions, that is, the end of what are now described as their outdated "privileges."

> When I started at university in the 1970s, things seemed to be going well. In '68, change even seemed to be on the cards. And then, with the arrival of programs of austerity, selection, evaluation – in short, everything that comes with the logic of competition – one quickly saw the limits of the model of the university. It was unable to resist . . . Today, it is a system under pressure . . . a context in which the monoculture of competition and evaluation is the rule – which has different effects across the disciplines, but somehow is killing them all, each in its own way.[17]

This is, par excellence, a matter for ecological analysis: "Competition can very easily accommodate weak, unreliable results if the effect of their being announced is to attract investors; if others pay the price for consequences that have not been taken into account or were not foreseen."[18] In the *knowledge economy* that is becoming more and more dominant, scientists are becoming more *dependent* on private interests. Let us make a note of this feature: dependency. It is etho-ecological by nature: "Ecological intelligence is an intelligence concerning the difference between interdependence and dependence, and it concerns how interdependence obliges us to act."[19]

As it happens, scientists have thus entered our new period in a weakened state: "In identifying science by means of consensual reasons – the objectivity, rationality, and method that are the privilege of the scientific mind – they have made themselves vulnerable to being called upon to produce an objective or rational approach to all the questions that the public or their funders, are interested in."[20] In 2006, in *La Vierge et le Neutrino*, she returned to this problem in detail, a book with the appropriate subtitle, *Scientists in Agony*, in the original French. She emphasizes that researchers' vulnerability may be related to their double game. They learn to conduct themselves differently, whether they are among colleagues or in front of the public:

> With her colleagues [in this book, Stengers uses "he" or "she" interchangeably] she will assemble, hesitate, and even, as the case may be [when she makes a breakthrough], dance. But when she addresses the "public," she speaks in the name of a reality that would, in itself, have the power to arm its representatives against illusion, a reality that would make of her a judge.[21]

Here we return to Stengers' proposition of practitioners presenting themselves with their requirements and obligations that would allow them to speak well of their work and of their successes. The scientists' *obligations* don't transform them into *judges* but engage them in an adventurous creation of relations with what they address. The catastrophe on the horizon is one where scientists think they are resisting by situating themselves "in a besieged fortress, faced with a world that is generally irrational and therefore hostile."[22] When, in France, the *Sauvons la recherche* [Save research] movement arose, Stengers kept asking: "save it from what?" From the "rise in irrationality" or its submission to private interests? It is the incapacity on the part of scientists – even if there are notable exceptions, like Barbara Van Dyck, who took part in tearing up GMO crops at the expense of her university job in Leuven, and to whom Stengers wrote a long homage[23]– to make alliances with the public that disarms them in their new situation: "Scientists are on the defensive, but they cannot 'chase away' these new protagonists [industrialists now asked to 'partner' with public research] because the financing of their research, like the 'worldly' extension that permits scientific [or techno-industrial] progress to be identified with human progress, depends on them."[24] "'Science' and 'Progress' are being turned back against researchers and rendering them doubly vulnerable: in relation to their traditional allies who now impose their own priorities upon them, and in relation to a 'public' that is beginning to break with the role of satisfied beneficiary that had been assigned to it."[25] And all this while the experimental sciences are considered a model, as a methodology – opposed to an adventure – that is sufficient for any domain it can be imported into in order to ensure the "advance of knowledge." Stengers wrote about this issue: "The idea of 'advance' may be handy for the successes in the experimental sciences, but it is a real poison for more 'terrestrial' research practices, ones that are dealing with crucial, but fragile, entanglements, that can be destroyed without even knowing it."[26]

Scientists know how to deal with "active" stakeholders (their colleagues, funding bodies), but they ask the "public" to just buy in. But for Stengers the question is how to "incit[e] them to envisage different relations with their milieus."[27] How can they become able to discard the comfortable motto of "science enlightening opinion"? How can they make themselves less vulnerable to techno-industrial predators? She writes, "It is a political problem, which means that the sciences are in need of stakeholders, like activists, who demand these other ways of doing science. This is why it is crucial for me that the activists do not denounce 'science' as such but incite scientists to betray their role."[28] When activists proclaim, "We are not defending nature, we are nature defending itself," this could be equally valid for researchers who require a fabric of interdependency. Activists could call them to recognize themselves "not as situated within an environment, but as belonging to this fabric, and thus as liable to destroy it. They [would] call for researchers to learn how to participate in the weaving of the fabric and accept that they are themselves woven in it."[29] As Latour says, the virus is not in our homes; we are the ones who are in the world created by the virus.

Stengers also highlights another change linked to the intrusion of Gaia:

From the point of view of what Whitehead called science, times have changed: physicists . . . can no longer make us forget this Earth, with its entangled histories in which they too participate, even if their laws make abstractions of them. Some figures with scientific credentials [Lynn Margulis, Rachel Carson] have given us a taste for a science that makes us understand entanglement, [in which] . . . [c]auses are everywhere, yet none of them is capable of defining its effects, "all other things being equal." In fact, on Earth, things are never equal, and only rarely can causes be defined independently of what they participate in. Even the notion of system now feels too reassuring.[30]

This is something that resonates perfectly with what Latour says here:

And to complicate the situation further, the scientific disciplines that have come together to develop these facts that have become so sturdy do not come from the prestigious sciences such as particle physics or mathematics; they come from a multitude of earth sciences whose certainties have been achieved not by some earth-shaking, foolproof demonstration but by the weaving together of thousands of tiny facts . . . a tissue of proofs that draw their robustness from the multiplicity of data, each piece of which remains obviously fragile.[31]

Fieldwork science has just taken precedence. And those that are specialists in it can certainly learn to conduct themselves differently from their colleagues confined in laboratories, formatted as they are by the new knowledge economy.

Latour also takes note of the change of era: "the modernization epic is over."

Everyone is aware that what happened in the process called "modernization," was something quite different from what they were telling themselves, that none of the schemas used to understand it – Enlightenment, progress, rationality, emancipation, humanism, not even the negative stories either – none of that was ever able to grasp, in an historical or anthropological fashion, what had happened over the last three centuries.[32]

We saw how for Stengers, the GMO issue was a foundational event. We know that "molecular biologists [claimed] that their strains of genetically modified plants could solve the problem of world hunger," batting away "the doubts of their colleagues who pointed to the socioeconomic reasons for famine, to social inequalities."³³ They were behaving like sleepwalkers. As for the activists, they knew how to distinguish GMOs in laboratory isolation from GMOs

> as an agricultural innovation. They understood that a fact may be solid but only from the point of view of trials that put its solidity to the test. It remains mute about everything that these laboratory trials demand to omit . . . In the case of the genetic modification of a living being, omission is not a project but the price for the effective verification of the modification. This is why the obtained facts have no legitimate claim to mute the questions associated with the consequences of this innovation beyond the laboratory. We need not reduce their objectivity to an imaginary social construction. We need only emphasize that [it] is intrinsically precarious . . . The only objective definitions that escape this precariousness are those that have been designed to rule on any territory whatsoever . . . [they] then occupy territory, as it were, and, like an occupying army, demand silence and submission.³⁴

This brings back the bifurcation of nature, having moved from a philosophical to a political register:

> bifurcation begins to operate as a veritable machine of government, distributing responsibilities in a binary and asymmetrical mode. On the one hand, the ceaseless bustle of beliefs and value judgments are considered arbitrary . . . On the other hand, objective definitions are attributed the power to bring humans into agreement, or if not, the power to silence them. Everywhere the same imperative prevails, which puts Science in the service of public order.³⁵

It is a matter of "keep[ing] the people silent," as Latour had said already.³⁶

Yet Latour will paint a more optimistic portrait of researchers in the Anthropocene era. The new situation will force them into it, and the opportunity has to be positively grasped. Concerned researchers are in fact no longer going to be able to distinguish *facts* and *values*, *matters of fact* and *matters of concern*: "In the current context, there is no alternative. A scientist has to appear cool, distant, indifferent, and disinterested. For several seconds, in suspense [Latour is telling the story of a theatrical fiction he participated in], Virginia explores other solutions, each one more calamitous than the one before." Then, "in a moment of inspiration and panic, she cries out against Ted [the character representing climate sceptics], whom the spectators are on the verge of driving out of the room: 'Go tell your masters that the scientists are on the warpath!'" And, she adds, addressing the same scientists:

> If your adversaries tell you that you are engaged in politics by taking yourselves as representatives of numerous neglected voices, for heaven's sake answer "Yes, of course!" If politics consists in representing the voices of the oppressed and the unknown, then we would all be in a much better situation if, instead of pretending that the others are the ones engaged in politics and that you are engaged "only in science," you recognized that you were also in fact trying to assemble another political body and to live in a coherent cosmos composed in a different way.[37]

Scientists are indeed the representatives of all these new agents whose roles cannot be denied by anyone. The "supposed unbridgeable distance between strict description and lively prescription that something should be done – without anyone telling us exactly what"[38] is reduced to nothingness. A threshold that is at the same time "juridical, scientific, moral, and political" is crossed. He sees cause for celebration in this:

> Earthbound scientists are fully incarnated creatures . . . They have enemies. Sometimes they win, sometimes they lose. They are of this world . . . They belong to the territory designated by their instruments. They have no qualms confessing the tragic existential drama in which they are engaged. They dare to say how afraid they are, and in their view such fear *increases* rather than *diminishes* the quality of their science . . . What for most people, including scientists, could be seen as a catastrophe – that scientists are now fully engaged in geopolitics – is what I take as the only tiny source of hope arriving to enlighten us in the current situation.[39]

So, there is no science war (à la Sokal), but a war of the worlds. Those who wanted to see Latour as an armchair critic will be disappointed: "There is indeed a war over the definition and control of the Earth, a war that pits – to dramatize the situation – humans living in the Holocene against Earthlings living in the Anthropocene."[40] As philosopher Patrice Maniglier puts it, Latour has become a "warlord."[41]

For her part, Stengers writes:

> Climate scientists, glaciologists, chemists, and others have done their work and they have also succeeded in making the alarm bells ring despite all the attempts to stifle them, imposing an "inconvenient truth" despite all the accusations that have been leveled against them, of having mixed up science and politics . . . They were able to resist because they knew that time was running out, and that it was not them but what they were addressing that was indeed mixing scientific and political issues.[42]

We can now come back to the question, which the Marxists call the "strategic" question, on which there has not been much discussion from those who have read Stengers and Latour. Stengers said that one can "inscribe oneself in Marx's heritage without necessarily being 'Marxist.'"[43] According to Stengers, Gaia obliges us to relinquish strategy:

> What has to be given up, at the moment of greatest need, really is what has most often served as the rudder for struggle – the difference to be made between what this struggle demands and what will become possible "afterwards," if capitalism is finally defeated. Naming Gaia, she who intrudes, signifies that *there is no afterwards*. It is a matter of learning to respond now, and notably by creating cooperative practices and relays with those whom Gaia's intrusion has already made think, imagine, and act.[44]

We are neither, to use Latour's formulation, before or after the Apocalypse, but "in an apocalyptic discourse *for the present time*."[45]

Stengers also warned that:

> the domination of abstraction is what is presupposed and realized by the process of commodification, when all concrete production is reduced to its exchange value in a regime of generalized equivalence, and when the living labor of human beings is evaluated as "labor force." . . . When they turn the capitalist redefinition of things and of social relations, or its contempt for the environment, into the privileged key for a reading of the modern epoch, both the militant revolutionaries and the protectors of nature rely on an abstraction that is in danger of being intolerant. Indeed, whoever says "key" says "mode of reading that is blind to its own selectivity," for the key designates its lock as the door's only relevant element.[46]

The Intrusion of Gaia 123

On this point – the way in which he avoids taking capitalism as the key to reading the situation – Stengers comes to Latour's defense, where a number of critics would have liked to separate them.

> This procedure of nominating [capitalism as the Enemy] was no doubt convenient for Marxist speculation, which bet on a possible newly baptized proletariat, whose powerful deployment would require a target against which to mobilize. But the witnesses of this possibility became strategists, and the price of this strategy was the disqualification of everything that might divert people from the imperatives of a mobilization front, one against one . . . Latour, for his part, proposes that political ecology should bet on multiple productions of "various imbroglios" that one has to try to "represent." This, of course, cannot be done without a "fight" because that is where the rapacity will be encountered. But the fight comes from a bet "for" and not a mobilization "against."[47]

And she makes her proximity clear:

> If I have allowed myself to paraphrase Bruno Latour's theses in this way, it is not because he needs to be defended, but because this type of wager (which I attribute to him) converges with the one that, in the work from which this word "cosmopolitics" emerged, I myself took risks in relation to so-called modern or scientific practices of knowledge.[48]

Latour often speaks of the "common world" to be composed and Stengers of the *possibilities* that speculative thought must cultivate against the *probable*.

In an interview, Stengers elaborated:

> I have certainly derived more from Marx than Latour has. Yet, even in the perspective where social relations might be radically modified, a certain number of serious problems that have been posed continue to do so . . . Should the revolution happen tomorrow, we will still have climate disorder on our hands, massive inequalities across different parts of the world, pollution, desertification, etc. . . . So I am not Marxist in the sense that I do not refer to a proletariat that has nothing to lose except its chains, or to theories of alienation that seem to presuppose that removing the alienation will reveal humans capable of thinking, enjoying their liberty, etc. . . . Alienation theory is our Christianity: our cult of the victim, our way of aligning *victim*

with *truth* . . . Alienation presupposes liberty through detachment. What interests me is humans inasmuch as they are *attached*, inasmuch as they have attachments that make them heterogeneous in relation to each other, and that make them interesting and important for each other, to the extent that, if one detaches them for whatever reason, they will be less than what they are.[49]

Words that Latour could have written: "The terms liberation, emancipation, laissez-faire or laissez-passer must no longer command automatic adherence by 'progressives.' In front of the Liberty flag, forever raised to guide the people, we would be well advised to carefully discriminate, among attachable things, those that will procure good and durable ties."[50]

In the sixth lesson in *Facing Gaia*, he takes this point up again from a different angle, what he calls the unfortunate secularization of the earthly paradise. Like Stengers, he distinguishes activists (a notion that relates to that of "speculation" and indetermination) from militants (relating to military language, as in "mobilization," for example). "John Dewey's whole political philosophy . . . consisted in managing to distinguish experimentation, linked to the practice of investigation, from the application of a truth. This is what makes it possible to distinguish activists from militants."[51] He also characterizes as militants those who are "definitively immunized against doubt." His other bugbear is the certainty of utopianism, unlike the uncertainty that characterizes ecology and its political practices. Stengers defends a similar point of view when she writes, "I have tried to avoid stories that try to put utopia to the fore. Please, don't tell us nice stories for dreamers; tell us about journeys of experimentation, failures, errors, successes, at the very least, apprenticeships."[52] And, again, "There is but one certainty: that the process of creation of possibilities must avoid utopianism like the plague."[53] According to Latour, materialists are often idealists whose utopias are disappointed by "matter." What one can do on Earth is "possible only under the conditions imposed by the passage of time. And thus slowly, with difficulty, with loss, aging, care, and concern."[54] To speak of "ecology, of the terrestrial world, about uncertainty or fear and trembling before the ongoing distribution of agency," this is what it is like to live in a metamorphic zone like Gaia. He underscores the danger that after "the failure of [their] projects," the Moderns "despise the world and violently reject matter as unfit to be transformed by Ideas."⁵⁵ Gaia "is *an injunction to rematerialize our belonging in the world* . . . by requiring the Moderns to start taking the *present* seriously at last, Gaia offers the only way to make them tremble once again with uncertainty about what they are, as well as about the epoch in which they live, and the ground on which they stand."⁵⁶

Politics cannot be a pedagogy aiming to hasten the arrival of a new world that can be described in the abstract, the paradise on Earth with which Marxists have often flirted (the project of an abundant society where even politics disappears). Did Marx not write that when "class distinctions have disappeared . . . the public power will lose its political character,"⁵⁷ which does not allow for political thinking in the face of Gaia at the time of the New Climatic Regime or in catastrophic times?

According to Latour, this is not without a relationship to religion:

> The promises of the beyond have been turned into utopias . . . If the history of the Moderns had consisted in moving from the abandonment of illusions about the beyond to the solid resources of the here below, it would have become wholly *attentive to the terrestrial*. But *for those who have immanentized Heaven, there is no longer any accessible Earth* . . . In other words, if they miss out on this world, these Moderns, their failure results not from an excess of materialism but, rather, from an overdose of ill-placed transcendence.⁵⁸

It is hard not to see how this compromises the communist project:

> But to realize here below the promise of the beyond inevitably means passing from a definition that could be called spiritual to a form of politics . . . From that moment on, poor politics, so impotent, so modest, so concrete, always so disappointing, was charged with the crushing weight of making the Kingdom of the Spirit realistic [which Latour, following the argument of Eric Vogelin, traces back to Joachim de Flore (1130–1202)].⁵⁹

We can go back to the fact that Latour rarely uses the word "capitalism." He often explains that it is the best way to make oneself powerless, that nothing is more disarming than this "big concept," which presents itself as already having described what in fact needs to be well described, while making the description more difficult at the same time. [Totalization] "renders its practitioners powerless in the face of the enemy, whom it endows with fantastic properties."[60] For Latour, the idea of capitalism ends up creating an overarching entity, leaving no place for effective footholds. Already, in 1994, he wrote:

> If you set yourself the task of following practices, objects, and instruments, you never again cross that abrupt threshold that should appear, according to earlier theory, between the level of "face-to-face" interaction and that of the social structure; between the "micro" and the "macro" . . . You do not have to choose your level of analysis at any given moment: just the direction of your effort and the amount you are willing to spend . . . *Social worlds remain flat at all points*, without there being any folding that might permit a passage from the "micro" to the "macro."[61]

Stengers wrote:

> It is in this way that I believe I understand Bruno Latour's special mode of presentation which has scandalized more than one reader of the *Politics of Nature* . . . [he] announces the possibility of calling anti-capitalism into question, and any reference to capital and a "spirit of capitalism," but proposes instead to combat monopolization by "certain groups" who, in the name of a science of economics, claim to be the only qualified representatives of the multiplicity of organizations and apparatuses that define market exchange . . . the struggle that Latour calls for will obviously not first involve a critique of the claims of economic science [. . .] but rather what American activists call *reclaiming*, concrete struggles against monopolies, wherever this becomes possible. But what is proposed is not giving in to fascination by a Being, Capital, whose name would redouble the power of "certain groups" by defining them as being in the service of something even more powerful than themselves.[62]

In *Capitalist Sorcery*, Stengers inherits some of Latour's distrust of "big concepts." She even made it a constraint: "[W]e are very interested in the calling into question of 'big concepts.' For example, the manner in which Bruno Latour attacks notions of 'Society' or of 'Science' or of 'the Scientific Spirit,' which serve to explain, whereas they are what should be explained."[63] She had this obligation in mind when she characterized capitalism by the way it *takes hold* and not first, abstractly, as a mode of production, thus refusing to make Marxism an "economic theory" (which many writers of Marxist "treatises" on economics have done, whereas Marx himself always presented his work as a critique of political economy). This was, in her view, the only way to account for it and to learn to oppose it. Stengers even wrote, "In contrast to Gaia, one ought to associate it [capitalism] instead with a power of a (maleficent) 'spiritual' type, a power that captures, segments, and redefines for its own purposes ever more dimensions of what makes up our reality, our lives, our practices, in its service."[64]

From the outset, taking up this question of "strategy," we characterized ourselves in *Capitalist Sorcery* as *depth sounders*, not *prophets*:

> We hesitated a lot before realizing that we would fail if we could be attributed the position of the person holding the map indicating the direction to take, the distance to cover, the usable paths and steps to follow. We do not know. We are neither prophets nor theorists . . . [Depth sounders'] knowledge stems from the experience of a past that tells of the danger of rivers, of their deceptive currents, of their seductive eddying . . . Sounders of the depths may well stay at the front of a ship, but they do not look into the distance. They cannot announce directions nor choose them . . . [they] should not invent words that are to be understood as beyond division, as if they were authorized by a transcendence in the presence of which everyone must kneel.[65]

This "we do not know" brings into existence "a trembling possibility where previously probability reigned."[66] Speculation! The anti-capitalism that interests Stengers relates to a "logic not of mobilization but of connection, and the production of connections implies a milieu that could be described as 'activated,' recalcitrant."[67] So there is no "strategy" or "theory" coming along to guide the action: "How can we become capable of fabricating something that concerns a future which we have no idea about, knowing that what we do or don't do today may not make any sense tomorrow, that it could poison or nourish those who are going to have to deal with it in a more direct way than us?"[68]

But the Great Divide has also had disastrous effects on some of Marx's descendants. One can always force the issue and take up the utopian thread in Marx (often in addition to Marx the economic theorist). This is not the choice that Stengers made, having preferred the more pragmatic Marx, that is, the one who pays attention to the consequences of his propositions. She is not worried about praising "opportunism":

> A very fine word "opportunism" since it designates the meaning, which is a force, of what is opportune, of what is suitable for each situation, the meaning of *this* "concrete" situation, accompanied by the halo of what can become possible. A word destroyed by those who wish for a theory or principles that guide action and guarantee choices without having to produce the force to think them.[69]

The notions of *alienation* and *consciousness raising*, i.e., consciousness free of all illusions, are typical of what reactivates the Great Divide within critical or Marxist thought. Inspired by *A Thousand Plateaus*, Stengers comes back to "itinerant creation" that implies "an ambulant people of relayers."[70] It corresponds to one way of doing [*faire*] politics (where the verb *faire* signifies once again "fabricate") that

> makes the need that minoritarian groups have of one another exist . . . The passage of relays implies not only holding but also giving. For the relay to be taken, it must be given, even if those who give know that they are not masters of what they give, that when a relay is taken it is not a matter of a simple translation but of a new creation. In this regard, Félix Guattari evoked a process of "existential catalysis," wherein each "creation" or reconquest is able to generate repercussions in the mode of the "yes, it's possible," able to arouse the appetite that will make another possibility exist elsewhere.[71]

So Stengers then proposes speaking of a *generative* process: "A generative practice can give rise to other possibilities that were not as such included in the beginning; it creates a sensitivity for other possibilities."[72] Here we are very close to what Anna Tsing brought to light in *Friction* when she related how activists borrowed from each other, from one country to another, learning to pick up on relays, adopting while transforming everything. It has always to do with apprenticeships.[73]

So this is indeed a continuation of what our two authors have taught us about the sciences and which has kept us so busy. The whole of the first part of this book was therefore not unrelated to what we are now presenting. Scientists know how to "relay" in certain situations. They have an *obligation*: "an experimenter will be obliged to recognize every objection as legitimate and welcome, on condition that this objection is 'competent,'" coming from the development of "communities of competent colleagues." We have seen how this relates to a politics of the sciences. But there is another side to the coin. Scientists have not generally learned how to take into account what is more directly concerned with what might indeed be obligations, but that bear on what lies *outside* of their collegial communities: "Scientists have not presented their successes as always selective and risky but as assured by a method, and what is more, a general method," the famous *scientific spirit* dear to Bachelard's heart. They have to learn how to give "an itinerant definition of what a success is, a definition that ought to reinvent itself whenever the situation changes . . . [which] would mean assembling not only 'competent' colleagues but also accepting that there will be other kinds of objections, coming from people not sharing the same obligations."[74]

This idea of people acting as "relays" came back to me when I read *After Lockdown*. Here Latour has dubbed one group of people the Menders,[75] in opposition to the Extractors. These Menders receive and give, and those who give do not know what those who receive will do with the gifts. They are like a cross between relayers and the American activist reclaimers. This is not about a "return to the past": what is mended will never be the same as the original. Things have to be taken up again and repaired!

Stengers suggests we engage with experiments that "those who know best" in France would treat with sarcasm and irony. She is interested in neo-pagan witches, and in particular Starhawk: "those who made the choice of calling themselves witches and activists have posed the hypothesis that to resist [capitalist sorcery] . . . imposed the rediscovery/reinvention of old resources, the destruction of which has probably contributed to our vulnerability."[76] They made the choice to challenge their disqualification, which ratified destruction in the name of progress, in order to become able to create new ways of living and struggling. They invented *empowerment* techniques that enabled them to resist at the most difficult moments, in particular in the United States when Reagan came to power, right through to today.

There are also of course the apprenticeship techniques that Latour experimented with in *After Lockdown*. What he suggests here is a devised arrangement that could also be characterized as an empowerment technique. The aim it to learn how to situate oneself in a way different from latitudes and longitudes: via what one *depends on*, through chains of dependency in which one is inscribed, and which one has to collectively learn what there is that is positive about them, or what is intolerable, and how they can be strengthened or changed. As Stengers says: "to heal and to reappropriate, to learn again and to struggle. Not to say 'it is ours' (compare the sad slogan 'everything is ours, nothing is theirs'), not to think ourselves victims, but to become capable once again of inhabiting the devastated zones of experience."[77] You can already hear the sniggering of those who have nothing but contempt for Latour's recipes, or the witches' ones . . . but, in both cases, the same questions do have to be relayed: "'And you, where do you draw your capacity to hold up and to act from? How do you manage to create the protections that the poisoned environment we all live in requires? What protects you from the vulnerability that the common enemy has continued to take advantage of? How do you do this? What have you learned?'"[78]

In Latour's case, what he needed to invent was a device that people could use not just to *localize*, but rather to be *situated in* in order to *feel* things.

> What do I depend on to subsist? What are the threats that weigh on that which provides for me; what confidence can I have in those who tell me about those threats; what can I do to protect myself against them; what are the aids I might find to help me extricate myself; who are the opponents I have to try to contain? . . . A territory will extend *as far as* the list of interactions with those we depend on – but no further . . . we learn to list *attachments* that force us to *take care of them* . . . what Isabelle Stengers calls *obligations*; the more precise your description becomes, the more it obligates you.[79]

What kind of device can best account for this new way of describing territory, from near to far?

> With Soheil Hajmirbaba, we tried to do this by drawing a large circle on the ground, oriented by an arrow, with a sign saying *more on*

one side and a sign saying *less* on the other . . . when you get near the middle, everyone gets a bit nervous: you have to make up your mind, and that's the hardest thing, you reveal yourself; you're going to talk about yourself or, better still, about what keeps you alive . . . about what threatens you, and, lastly, about what you are doing or not doing to counter that threat . . . every time you're about to name out loud one of the entities on your list, someone in the gathering comes "to play" that "role," and it's up to you to place this character on this sort of compass – or to move them around as your short narrative evolves. The amazing result of this little enactment is that you're soon surrounded by a small assembly, which nonetheless represents your most personal situation, in front of the other participants. Little by little, you're giving shape to one of these holobionts that seemed, till now, so hard to represent. A woman in the group sums it up in one phrase: "I've repopulated myself!"[80]

Obviously, there remains an open question: will Latour's idea be relayed by other Menders getting busy across multiple sites?

If *localize* and *situate* have to be distinguished, it is for a reason that Latour has consistently interrogated throughout his *oeuvre*, the question of *materiality*. What world are we living in? Is it that of the economy, as conceived by the Moderns, whether

La Mégisserie, Saint-Junien, 1 February 2020. Launch of the "Où attérir?" ateliers. Photography: Nicholas Laureau.

defenders or enemies of capitalism? This is what Latour is asking us to renounce. For too long, we have thought that the economy was our infrastructure. Not only that, but it has to have a *top* and a *bottom*, and no *imbroglios*! Not only that, you have to believe in it! From 1999, in his *Politics of Nature*, Latour drew on the work of Michel Callon, who had shown that markets are never given, and have nothing "natural" about them, but are the result of long formatting of both supply and demand.[81] He wrote:

> There is no such thing as an economy, just as there is no such thing as a *Homo oeconomicus*, but there is indeed a progressive economization of relations. We do not find, at the bottom, an economic infrastructure that the economists, situated above, would study: the economizers (in the broad sense of the term, which has to include accounting systems and modeling scenarios, mathematicians, marketing specialists, and statisticians) *performed* [*formattent*] the collective by stabilizing the relations between humans and nonhumans ... As soon as we have extirpated economics both from our heads and from the world in order to reduce it to a set of specific and uncertain procedures ... political economics loses its venom and stops competing with political ecology.[82]

This does not mean that the economists' calculations are useless, but they have to be put back in the right place: "Dangerous as infrastructure, economics becomes indispensable as documentation and calculation, as secretion of a paper trail, as modelization."[83]

When the French government reauthorizes neonicotinoids, it does so in the name of economy versus ecology: the beetroot that is grown, the farmers who grow it, the sugar factories, and the workers who work in them are part of the economy; on the other hand, the bees that neonicotinoids kill and the farmland ravaged by pesticides are not! What a strange infrastructure! On the one hand, what *counts*, on the other, what we are asked to *neglect*:

> When the FNSEA ... the French farmers' union, lays siege to the Ministry so that it will re-authorize the use of bee-killing pesticides in order, they say, "to save the French beet-sugar industry," we should not necessarily see this as having an "economic dimension," if what we mean by that phrase is a calculation of indisputable interests that ought automatically to save forty thousand jobs and so many billions

of euros. There has been a prior distribution of lifeforms, each one of which merits questioning . . . This doesn't amount to whingeing until we get other preoccupations put above the Economy, concerns that are supposedly "more noble," "more human," "more moral," or "more social." Quite the opposite: it means really taking stock of the fact that it's high time we delved *further down* by becoming more realistic, more pragmatic, more materialistic.[84]

This example indeed shows that the economy is never a *given* but always the result of delimitation work. The economy should no longer have the power to put a brutal end to *calculemus*. "The thing that terrifies the 'chief clerks' is that, in quitting the economy, terrestrials are really going home and returning to ordinary experience."[85]

It was Tarde – for whom the whole "is always *inferior* and *smaller* than the parts" – who helped Latour think through the importance of the distinction between the economy and economics as a discipline (the same word is used for *economy* and *economics* in French).

> In order to understand what makes Tarde so innovative in economics, it is necessary to fully grasp the innovation he brings to sociology. The idea, made famous by Polanyi, of an "embeddedness" of that which is economic in that which is social had the great impracticality of assuming the prior existence of society. We can understand, then, that the theoretical gain could not be very great: in passing from economism to economic sociology, all that was happening was shifting an already-established structure – the infrastructure and its laws – to another structure, it, too, already in place: society and its laws. Of course, we learned a lot about the "extra-economic" factors of contracts, of trades and of tastes, but it was, in a way, to move from one structure to another. Yet the "involution" Tarde proposes of all the laws of a structure in the swarming of monads had the drastic consequence of dissolving all structures – that of the pure and perfect market, of course, but also those of the social world which are accepted by sociologists like Durkheim and his disciples.[86]

Latour has a radical proposal: "it's not about making a fresh attack on economics; it's about abandoning it altogether as a description of the relationships that lifeforms maintain with each other. The Economy casts its spell, but we need to learn to exorcise it."[87] The economy is not an "infrastructure" for Latour because there is no infrastructure. The "matter" that would make up the economy is idealistic because it has gotten rid of all the worries that characterize Gaia, those concerned with *engendering*. Because the entanglement of all the beings constituting Gaia has no infrastructure at its base. So we need other concepts, other ways of *orienting* oneself in order to do politics. So this is where Latour introduces the notion of *engendering*, or *subsistence*. And Stengers, with Whitehead, spoke of *persistence* or *endurance*.[88] With the beetroot example, there are only beings who want to subsist, who have engenderment concerns: farmers, workers, sugar factory bosses, bees, soils modified by insecticides. There are no different levels, no transcendence, no bird's-eye view, no privileges or underpinnings, just the one plane of immanence whose name is Gaia. So there is an apprenticeship to be undertaken: *de-economize* the world in order to *ecologize* it.

This resonates perfectly with Stengers' position: "Those *ultimately responsible* have had their day ... there is now only a multiplicity of modes of existence, biological and non-biological, which need interweaving environments that give them meaning and where they make sense."[89] We have to *move out* of the Economy in order to *move in* to the ecology, and, from that point on, reconsider the idea of property. How can property rights be appropriate in this interdependent world? The world of economics depends on a material world that it does its best to ignore.

But what becomes of the social classes as defined by the former structure (infrastructure) of the economy? Latour opens up a new work site, which is ongoing.

> In the 1960s, even though they were already very contested, people situated themselves in these class structures. They organized the landscape. We must therefore reconstitute classes, no longer according to the positions of individuals in the production process that characterized industrial societies, but according to the territories on which they depend in order to satisfy their vital needs. This is what I call geo-social classes, the point of which is to allow the emergence of the class conflicts necessary for the organization of a future political course. It is necessary to build a geo-social class consciousness that allows each person to understand that he or she is in a struggle against other classes that are in the process of screwing up his or her conditions of existence by living "off the ground" through their level of consumption of oil and natural resources, the

degradation of the ecosystems implied by their way of life, etc. This will only happen if we can name what we disagree about, i.e., create lines of conflict.[90]

The Gaia moment is also the Whitehead one; the modes of abstraction and what they negate, as they feed into the bifurcation of nature, are perhaps more appropriate for laws of the Universe, which physicists like to say would also be recognized by extraterrestrials. But they do not allow us to apprehend the "critical zone." There, everything is deeply entrenched in its characteristic entanglements of living and non-living entities.

Latour goes on to write, after seeing a demonstration of high school students, "I was very touched to see placards that almost invariably alluded to the possibility of enduring, perpetuating oneself, and not just for humans, but all earthly things, glaciers, forests and, of course, animals."[91] And, for the economy, it is also the notion of production that loses its relevance:

> Production begins with a given world made up of resources. But when we speak about engendering we locate ourselves at an earlier level, and, above all, *at an earlier stage*: what is it that allows these resources, and these people who live off them, and these worlds in which they are born, to continue to exist? . . . the engendering crisis is affecting everything: all institutions, all peoples. The engendering hiatus has opened wide: oil, climate, insects, States, languages, children, etc.[92]

One can see the abyss that is going to open under the feet of quite a few (Marxist) militants . . .

This reminds us of what he wrote in *Irreductions*, almost forty years ago:

> I don't like the flamboyance of princes. I do not like power that burns far beyond the networks that fuel it . . . reduce the reductionists, that is what I want, escort the powers back to the galleries, networks, gestures, works they use to extend themselves. I want to avoid granting them the potency that lets them dominate even in empty places they have never been . . . There is no longer an above and a below. Nothing can be placed in a hierarchy. The activity of those of rank is made transparent and occupies little space. There is no more totality and therefore no more leftovers. Places are precise, with no excess or weakness. It seems to me that life is better this way.[93]

In an *ecologized*, rather than an *economized*, world, the Menders are facing off against the Extractors:

> The old scenography relied on the Economy since it was through their position in the "system of production" that injustices were spotted. But in these strange new battles, the Economy is no more than a superficial veil, and we're no longer dealing with production. What's at issue now are engendering practices and the possibility, or not, of maintaining, continuing, even ramping up the livability conditions of lifeforms which, by their action, maintain the very envelope inside which history never ceases to unfold ... Whence the obligation to reconstitute the nature of the soil particle by particle, yes, to mend, when every detail of critical zones is a world in itself that involves us and forces obligations upon us.[94]

Conclusion: Composing a Common World... During the Meltdown

"Dancing in their laboratories." I'd like to pick up on Stengers' description of the joy that can seize laboratory researchers at the moment they realize that their apparatus has given them the power to speak in the name of the thing they were investigating. But the joy that is important to her is also that of scientists who have been able to invent other ways of doing science than those specific to the theoretical-experimental laboratory. It is Shirley Strum's joyous energy among her baboons, or Lynn Margulis with her holobionts, as well as Vinciane Despret discussing birdsong, when she shares ethologists' hesitations, hypotheses, and knowledges. The opposite of joy is not just sadness; it is also boredom. I was not bored for a second, as I reread the body of Latour and Stengers' writings, and I hope those who come to the end of this little book will have the same feeling. Perhaps because one has a clear sense that they themselves are never bored as they compose. As Stengers suggests:

> To refuse to be bored when writing also means trusting the reader, that is, trusting the becoming, undecidable in any case, that will decide "how" the reader will read you on the basis of what demands, that is, of what experiences. The readers, for their part, are thus called upon to leave behind the status of an ignoramus, faced by an author who says what it is appropriate to think. They have to head, in their turn, "down the road," a road on which thought stumbles at every step and must invent itself all over again.[1]

Everyone will have been able to see that I have written this book more as an *operator* than as an author: "Whatever your intentions, if your objections force you to think, to discover what the Idea designating me as an operator requires, to follow the calculation to the point where it becomes more demanding, when concepts already created have to undergo terrible contortions in order to articulate with new elements, you are a friend."[2] If the operation works, it means that the reader has welcomed the induction, in the sense that "induction is not a suggestion, for it proposes a taking up of the baton but does not say how to do it. It modifies the sensitivity."[3]

Stengers speaks of the joy of those who "are in their element," that is, who have learned *how to learn from* what they study:

> This is the case in genetics, when Barbara McClintock, busy and precise, nevertheless affirms the importance of "listening to what corn has to say." It is the case in embryology, when Albert Dalcq is moved by the fact that the embryo's response to his experimental questions has "all the surprise and charm one can find in the answer of an intelligent interlocutor." Also, in evolutionary biology, when Stephen Gould opposes to the nasty ugliness of all-terrain selectivism the joy of a biology that would learn how to understand how, when it comes to living beings, "the world outside passes through a boundary into organic vitality within." And in ethology, when meticulous study is joined to a link of respect, or even love, or else amazement, or again jubilation.[4]

It can also be the case with a plumber, or an activist, or also a mathematician, or even animals who all create/appropriate worlds, worlds that we have to learn to respect (Stengers speaks of *politeness*, following Vinciane Despret). This is also the meaning that should be given to speculative philosophy: addressing oneself not to those Stengers calls "anyones"[5] but to people "attached" to a world that one must be careful not to ravage and into which "they admit no free trespassing."[6]

These ways of being "in one's element" could be like Latour's modes of existence, to each of which he attributes its dignity, a dignity that depends on its capacity not to occupy all the space, which was also one of Whitehead's preoccupations. The passion of the novelist is not that of the physicist, or the ethologist, sociologist, or priest, but none can claim to be able to deconstruct the others, to speak the truth for all.

Conclusion

There is no reason to seek to re-enchant the sciences (which would be an inadvertent way of bifurcating nature). They are neither cold nor inhuman when those who practice them make an adventure of it, when one knows how to distinguish those "that are passionate . . . from the others that are sad and mind-numbingly boring."[7] Good sciences do not *reduce*, but make new actors *proliferate*. All of Latour's work, and all of Stengers', has tried to disrupt the assimilation of the sciences to Progress.

It is not by chance that *reason* is initially part of the same declension as *reasonable*. It seems to me this is what the works of Latour and Stengers avoid. Those who present themselves as revolutionaries but who continue to vaunt an overarching Reason, a scientific Spirit, fall back into routine, repetitiveness, and imitation: they quickly become reasonable . . . disillusioned, lose any sense of defiance, of rebelliousness, of adventure (which they quickly assimilate to opportunism), in short, any imagination. And this applies to both science and politics; thinking about the former is always also thinking about the latter. We can hear the panicked asking, "They are not going to take the economy away from us, are they?" The partisans of Reason are always prepared to take the risk of becoming witch-hunters. Shouldn't one always be ready to take on the "rising tide of irrationality"? Condemn common sense "which is always wrong," to use Bachelard's phrase again? Latour, who likes to remind people he comes from a bourgeois family from Burgundy, is sometimes surprised to hear us say he is the most revolutionary of us all . . .

I have spent a lot of time in the pharmaceutical industry. Most of those who visit the research centers of the big pharmaceutical laboratories are often struck, like myself, by the sadness that permeates everything. Everything is routine. There is no dancing in laboratories where operations are organized, like in a factory, on a production line. Even the bosses were concerned that sterility was the end product, and they turned to outside start-ups to find some inventive dynamism. But they now also know that by absorbing them, by subjecting them, by integrating them into their own logic, they quickly lose their energy. A squaring of the circle!

And those who try to really engage political questions are always pressured to stop doing it and submit in the name of rationality. This imperative resurfaces with every crisis. Politics then disappears in favor of another, pedagogical, task: making people rational or more aware of insuperable necessities. Reason claims to pre-empt everything. "*Gorgias* . . . is obviously the founding text of this contempt for the political. But it's extraordinary, all the same, to see Socrates claiming he will win the argument with Callicles by demonstrating a mathematical theorem to a handsome young man while Callicles concerns himself with the agora."[8] When the sciences are confused with Reason, the scientific Spirit, and Progress, the aim is always to silence the screaming in the streets.

So now, as we are reaching the end of this "speculative gesture," this "entangled flight," something strikes us. Latour always takes the same tack: "this is a poor description of us," "we are not what we claim to be," "the descriptive work has to be started over." So, if we really are Moderns, it is not because of the claims we are making about ourselves. It could even be that this false description is the only remaining characteristic that allows for our definition, and that it constitutes a war machine. As the phrase, that is attributed to North American Indians, and which Latour likes to repeat, goes: "White man speaks with forked tongue."[9] For her part, Stengers always begins with the operations that have mutilated us, that have dismantled practices and knowledges and have impoverished the world, depriving it of resources that are, by definition, rare and precious. We have thus taken ourselves to be the "thinking head of humanity." We have thus multiplied and repeatedly renovated the operations of the Great Divide; between those who know and those who believe, between those who want to destroy illusions and those who resist. Latour and Stengers have the same enemy: ideas of Progress and Reason that both justify deceptive descriptions and the calls to give them up. And they both have something else in common,

> [a] "machinic" becoming in the sense of Deleuze and Guattari, meaning that thinkers can produce this thought only because they have themselves become a piece, or a cog, in what has captured them, much more than they have created it. Thought is then no longer the exercise of a right but becomes an "art of consequences," consequences that leap from one domain to another, or, more precisely, that make interstices zigzag where a homogeneous right had seemed to reign, and make connections proliferate where a *"this has nothing to do with that"* had prevailed.[10]

One philosopher, with whom many readers would have been unfamiliar (because of his reputed difficulty), has stayed with us on this journey because Latour and Stengers were able to make him indispensable: Whitehead. Stengers notes, in 2020:

> Indeed, I reprised Whitehead in a manner that is not "mine." "My" reprise was activated by the situation that is ours today. It might be said that the question of the decline of our civilization [Whitehead's diagnostic] has given way to the question of its collapse. Cracking and grinding can be heard that indicate the calving of ice sheets. Grounds once considered secure are dislocating.[11]

"To relay Whitehead . . . what I 'need to understand' is the difficulty that I have, the 'civilized' me, with letting myself be effectively affected, touched, concerned by this world that our modes of appropriation have ravaged and are still in the process of destroying."[12]

Latour also draws from Whitehead and his "reparative surgery." It was Whitehead who enabled him to think, over and above the bifurcation of nature, that "Newton has 'happened' to gravity, and Pasteur has 'happened' to the microbes."[13] There isn't a knowing "I" on one side and a knowable object on the other, with the passage from the one to the other being (as epistemology would have it) via a "bridge over the abyss." Latour put the Grand Divide, the opposition between knowing subject and thing to be known, at the center of his work. With him, we learned to replace it with a "chain of experiences," which is about the world itself, as Gaia has taught us, as well as about the process of knowledge acquisition. "It is the fact that 'occurs,' that emerges, and that, so to speak, offers you a (partially) new mind endowed with a (partially) new objectivity."[14]

For Latour, as for Stengers, Whitehead is one of those rare philosophers who renewed the philosophy of science and society while absorbing the shock of Darwinism and the theory of relativity. He constructed concepts such as "feelings," which are required and not explained by any constitution of a "thinking subject." This brings us back to speech impedimenta, those "how shall I put this . . ." with which I introduced this book.

We have seen that the question of diplomacy takes up quite a lot of room in the exchanges between Stengers and Latour. We have to come back to it in conclusion because it is a way of taking up the question of the Great Divide with which we opened this book. Diplomacy is not a call for general goodwill or tolerance, or for placing war and destruction on the same footing as risks that have to be run. It is about us, the westerners, in the first place. It is a proposal to "civilize" us, the Moderns. How can a common world be composed? According to Stengers, speculative philosophy "obliges philosophers not to invoke any habendum allowing them to eliminate, forget, treat as an exception, or disqualify an element of experience."[15] The slightest murmurs must be heard. The philosopher should "accept a constraint to deliberately abstain from the great resource of philosophy, 'explaining away,' that is, by eliminative judgment."[16] We have seen that neither Latour nor Stengers denounce the sciences. On the contrary, what they love about them is what makes for their specificity, their rarity. This is why "the experiences involved in the production of relevant measures, the creation of criteria, and the differentiation between the illusory and the legitimate must [not] be denounced. However, for the 'necessity' proper to the speculative system to have any sense, they must lose their privileged scope."[17] The wager is one of taking as the prerequisite for peace "not [a] purification that subtracts but a creation that constructs, that adds and complicates."[18] This cannot be achieved "by some process of conversion," but must be fashioned "actively, intelligently."[19]

"Needing people to think" is a phrase that Stengers borrows from Gilles Deleuze, which for him related to the difference to be made between left and right.[20] She also proposed picking up the idea of an *ecosophy*, or the wisdom of the middle, from Félix Guattari.[21] The "wisdom of the milieu" is a phrase that could be quite suitable for Latour and his attempts at good descriptions of what we are attached to and what we hold dear. However, what is being destroyed in these catastrophic times is not just what allows humans to inhabit the Earth, it is all the knowledges, seized as practices, that do not correspond to the rapacious processes of abstraction associated with progress. Hence the importance of what Stengers called "generative apparatuses," that generate and foster the making of sense in common. Kant's famous injunction, "Have the courage to use your own understanding" commits the children of the Enlightenment to behave as active subjects, wielding the weapon of reason. Here it becomes "dare to taste," to let yourself touch, to let yourself be transformed by others so that a composition can be generated that belongs to no one: "'dare' does not signify the flattering heroism of confrontation or emancipation, reason liberating itself from the yoke of seductively reassuring illusions . . . 'dare to taste' means instead 'dare to begin.' If you wish to activate processes of discernment through which will be generated a 'knowing how' to appreciate will be generated, dare to let yourself be touched."[22] When there are generative apparatuses, the "feeling together" that they activate is not the result of relations of forces, but is obtained, generated, for each of us, with others and thanks to others. One must not forget, Stengers remarks, that "American activists adopted the practice of decision making through consensus from Quaker activists, precisely because it generated decisions that had to hold up when put to the test."[23]

Stengers also suggests rehabilitating the idea of *commerce*:

> "[I]s there a means of having commerce here, with these people?" is without doubt one of the oldest human questions, and the true art of commerce is the art of negotiation. It is an art that extends well beyond the domain of commodities . . . In many traditions, to look after sick people is to know how to negotiate with invisible powers, how to have *commerce* with them . . . We might say that in the two cases, commerce of goods and commerce of ideas, it is the slippery slope of what economists call the "minimization of the transaction cost" that must be warded off. The art of commerce is "costly" in both time and patience. But it is what makes traders exist as obliged by a relation whose existence depends on them, as operating the co-production both of themselves in relation and of what links them together.[24]

Let us hope that this book will be useful for those who will make the emerging tradition of Latour, or that of Stengers, bear fruit in their domains. One may well need to make use of both, and of the ways in which they have been able to have "commerce." In this conclusion, I would like to pick up on what Stengers says of Donna Haraway's string games:

> Several players are needed for this game; one does not play it alone; one does not come up with ideas oneself. At each stage of the game, a string figure is held out by one player, which serves as motif for the next player who takes it up. The first player lets go the figure

she held through the vey move when the next player takes over it, producing a new figure. Figures are successively made and unmade. The player who holds out a figure remains passive while another responds to the offer, engaging with the interlaced strings to deploy a new figure. Each new figure is at once deployed and exposed, open to serving as motif for another player.[25]

The alert reader has perhaps noticed that the first chapter was almost exclusively devoted to Latour, while the second was on Stengers. From then on, things interlaced more and more such that they become totally interwoven in the last. I dare to hope that it might often be difficult to attribute one phrase or another to one of them, had the name of the author not always been carefully cited. This is not a stylist effect, and I would certainly not have bet on it before launching on the adventure of this book. This is just one of the things that happened. So I found myself seized by the passion that is also a state of uncertainty. It is what is conveyed, as we said right at the beginning of the book, by Stengers' "how shall I put this . . ." Speech impedimenta! They can fail at any moment. For my part, it was a matter of prolonging an "induction which had already contributed to the modification of [my] own relationship to thought."[26] More than ever, I felt the force of this phrase from Deleuze cited by Stengers: "The imperatives of the questions that traverse us do not emanate from the I, which is not even there to listen to them."[27]

An entangled flight, as I said. It is time to come in to land:

> So, you've landed, you've crashed, you've extricated yourself from ground zero, you're advancing, masked, your voice barely audible: like Gregor's [the character in *Metamorphosis*], like mine, it's a sort of mumbling. "Where am I?" What to do? Go straight ahead, as Descartes advised those lost in a forest? No! You should scatter as much as you can, fan out, explore all your capacities for survival, conspire, as hard as you can, with the agencies that have made the places you've landed on habitable. Under the canopy of the heavens, now heavy again, other humans mingled with other materials form other peoples with other living things. They are freeing themselves at last. They're coming out of lockdown. They're being metamorphosed.[28]

How better to finish than by citing a few lines written by Souriau at the end of his book, repeated by both Stengers and

Latour in their only joint text: "[W]e can tell of a song, which sounds beyond existence with the power of supernatural speech, and which may be able to cause even the gods in their interworlds to feel a yearning for the 'to exist' – as well as the longing to come down here by our sides, as our companions and our guides."[29]

Notes

Introduction: Speech Impediments

1 "It is a way of talking about the way in which things depend on each other. And the act of weaving is an activity that creates dependence, passing over and under." Stengers, *Activer les possibles*, Noville-sur-Méhaigne: Éditions Esperluète, 2018, p. 95.
2 "What Deleuze sometimes calls 'friendship' doesn't have much to do with a friendship between persons, it's much more the knowledge that with this person, it's not necessary to explain oneself too much and that an 'exchange' might be possible, like an entangled double flight." Isabelle Stengers, *La Vierge et le Neutrino: Les scientifiques dans la tourmente*, Les Empêcheurs de penser en rond series, Paris: Seuil: 2006, p. 161. English translation: *The Virgin Mary and the Neutrino: Reality in Trouble*, trans. Andrew Goffey, Durham: Duke University Press, 2022 (in press). Goffey renders the titular phrase as "an entangled double flight." Elsewhere, Stengers writes: "Guattari has a beautiful phrase: There is a relationship between philosophy and friendship. And so, the modes of working with philosophers take place in a strange way, this friendship . . . but not in the sense of old mates who meet up and slap each other's backs, etc., but it is true that it is a space of practices that are quite distinct from the space of scientific practices." Isabelle Stengers, "Discipline et interdiscipline: la philosophie de 'l'écologie des pratiques' interrogée," *NSS* 8(3) (2000): 59–63.
3 Concerning the many encounters scattered along Latour's journey,

one can refer to Gerard de Vries, *Bruno Latour*, Cambridge: Polity, Key Contemporary Thinkers, 2016. But this book ignores all of Stengers, and as such it is unsatisfactory! There is not yet an equivalent book on Stengers.

4 Bruno Latour, *The Politics of Nature: How to Bring the Sciences into Democracy*, trans. Catherine Porter, Cambridge, MA: Harvard University Press, 2004. In the English edition, there are three dedicatees, including Stengers.
5 Isabelle Stengers, *The Invention of Modern Science*, trans. Daniel Smith, (Theory Out of Bounds, 19), MN: University of Minnesota Press, 2000.
6 Isabelle Stengers, *Au temps des catastrophes. Résister à la barbarie qui vient*, Les Empêcheurs de penser en rond series, Paris: La Découverte, 2009. English translation: *In Catastrophic Times: Resisting the Coming Barbarism*, trans. Andrew Goffey, London: Open Humanities Press, 2015.
7 Bruno Latour, *Cogitamus: Six lettres sur les humanités scientifiques*, Les Empêcheurs de penser en rond series, Paris: La Découverte, 2010; Bruno Latour, *Enquête sur les modes d'existence: une anthropologie des modernes*, Paris: Seuil, 2012. English translation: *An Inquiry into Modes of Existence: An Anthropology of the Moderns*, trans. Catherine Porter, Cambridge, MA: Harvard University Press, 2013.
8 Tobie Nathan, Isabelle Stengers, and Lucien Hounkpatin, *La Damnation de Freud* (a play in four acts), Les Empêcheurs de penser en rond series, Paris: La Découverte, 1997. Isabelle Stengers, *La Guerre des sciences aura-t-elle lieu? Scientifiction*, Les Empêcheurs de penser en rond series, Paris: Seuil, 2001.
9 Isabelle Stengers, *Penser avec Whitehead. Une libre et sauvage création de concepts*, Paris: Seuil, 2002. English translation: *Thinking with Whitehead: A Free and Wild Creation of Concepts*, trans. Michael Chase, foreword by Bruno Latour, Cambridge, MA: Harvard University Press, 2014, p. 463. Translation modified.
10 Stengers, *Thinking with Whitehead*, pp. 228–9. Translation modified.
11 Didier Debaise, "Les âmes du monde," in Fleur Courtois-L'Heureux and Aline Wiame, *Étienne Souriau: une ontologie de l'instauration*, Paris: Vrin, 2015, pp. 111–29.
12 Latour, *The Politics of Nature*, pp. 249–50.
13 Latour, *The Politics of Nature*, p. 237.
14 Bruno Latour, *Jubiler ou les tourments de la parole religieuse*, Les Empêcheurs de penser en rond series, Paris: La Découverte, 2013 (2002). English translation, *Rejoicing: or the Torments of Religious Speech*, trans. Julie Rose, Cambridge, UK: Polity, 2013, p. 5. Translation modified.

15 Stengers, *Thinking with Whitehead*, p. 464.
16 Stengers, *Thinking with Whitehead*, pp. 307–8. Translation modified.
17 Stengers, *Thinking with Whitehead*, p. 404. Translation modified.
18 Stengers, *Thinking with Whitehead*, p. 73.
19 This passage only appears in the original French edition, *Penser avec Whitehead*, p. 392 [Trans].
20 Bruno Latour, "La connaissance est-elle un mode d'existence? Rencontre au muséum de James, Fleck et Whitehead avec des fossiles de chevaux," in Didier Debaise (ed.), *Vie et expérimentation: Peirce, James, Dewey*, Paris: Vrin, 2007, pp. 17–43. Translation: "A Textbook Case Revisited: Knowledge as Mode of Existence," in E. Hackett, O. Amsterdamska, M. Lynch, and J. Wacjman (eds), *The Handbook of Science and Technology Studies*, 3rd edn, Cambridge, MA: MIT Press, pp. 83–112. References in this book are to the more readily available text on Bruno Latour's website (http://www.bruno-latour.fr/node/49.html) p. 12 [Trans.].
21 Stengers, *Thinking with Whitehead*, pp. 464, 465.
22 Stengers, *Thinking with Whitehead*, p. 518.
23 Isabelle Stengers, *L'Hypnose entre magie et science*, Les Empêcheurs de penser en rond series. Paris: Seuil, 2002, p. 140.
24 Stengers, *Thinking with Whitehead*, p. 462.
25 Isabelle Stengers, "Que vas-tu faire de moi?" in Fleur Courtois-L'Heureux and Aline Wiame (eds), *Étienne Souriau*, op. cit., pp. 63–85. Here it is interesting to pick up on Whitehead's terms. "Descartes in his own philosophy conceives the thinker as creating the occasional thought. The philosophy of organism [i.e., Whitehead's] inverts the order, and conceives the thought as a constituent operation in the creation of the occasional thinker. The thinker is the final end whereby there is the thought." Alfred North Whitehead, *Process and Reality: An Essay in Cosmology*, ed. D. R. Griffin and D. W. Sherburne, New York: Free Press, 1978 (1929), p. 151. This was discussed by Didier Debaise in "Alfred North Whitehead. Les sujets possessifs," in Didier Debaise (ed.), *Philosophie des possessions*, Dijon: Les Presses du réel, 2011, pp. 233–51. It is helpful to cite this passage as well: "The philosophy of organism [i.e., Whitehead's] is the inversion of Kant's philosophy. *The Critique of Pure Reason describes* the process by which subjective data pass into the appearance of an objective world. The philosophy of organism seeks to describe how objective data pass into subjective satisfaction, and how order in the objective data provides intensity in the subjective satisfaction. For Kant, the world emerges from the subject; for the philosophy of organism, the subject emerges from the world . . ." (p. 88).

26 "Tracking" in the sense highlighted by Baptiste Morizot, when a human "postulates that there are things to translate and . . . tries to learn. With this kind of attention, one is always in the process of garnering signs, always in the process of forging links, noting flashes of strangeness, and thinking up stories to make them comprehensible . . ." Baptiste Morizot, *Manières d'être vivant*, Arles: Actes Sud, 2020, p. 139.
27 Thinking in a speculative way means "intensifying the importance of an experience to the highest degree." See Didier Debaise, "L'intensification de l'expérience," in Didier Debaise and Isabelle Stengers (eds), *Gestes spéculatifs*, Dijon: Les Presses du réel, 2015, p. 112.
28 Stengers, *Thinking with Whitehead*, pp. 215–16, citing Étienne Souriau, "Du mode d'existence de l'œuvre à faire," *Bulletin de la Société française de philosophie*, February 25, 1956, pp. 4–24. English translation: *The Different Modes of Existence, Followed by, Of the Mode of Existence of the Work to be Made*, Introduction by Isabelle Stengers and Bruno Latour, trans. Erik Beranek and Tim Howles, Minneapolis, MN: Univocal, 2015.
29 Bruno Latour, "Reflections on Etienne Souriau's *Les différents modes d'existence*," trans. Stephen Muecke, in Graham Harman, Levi Bryant, and Nick Srnicek (eds), *The Speculative Turn*, Melbourne: re.press, 2011, pp. 304–33, p. 311.
30 Isabelle Stengers, *Réactiver le sens commun: Lecture de Whitehead en temps de débâcle*, Les Empêcheurs de penser en rond series, Paris: La Découverte, 2002, p. 165. English translation by T. Lamarre: *Making Sense in Common: A Reading of Whitehead in Times of Collapse*, Minneapolis: University of Minnesota Press, 2022, p. 150. The idea of a middle voice, contrasting with active and passive, doesn't exist in French, unlike many other languages, for instance Greek. See Émile Benveniste, *Problems in General Linguistics*, trans. M. E. Meek, Miami: University of Miami Press, 1971, pp. 145–52.
31 Stengers, *Making Sense in Common*, p. 151.
32 It is on this very question that Whitehead is in debt to William James. "Beginning with the *Principles*, James will therefore introduce arguments that will be picked up again in his famous article 'Does "Consciousness" Exist?' of which Whitehead, on the occasion of its publication in 1904, wrote that it 'inaugurated a new era in philosophy' succeeding that which was opened up by the *Discourse on Method*." Isabelle Stengers, "William James. Naturalisme et pragmatisme au fil de la question de la possession," in Didier Debaise (ed.), *Philosophie des possessions*, op. cit., pp. 35–69. This article deals in detail with the question of the "I," which is only briefly examined here.

33 Bruno Latour, "Pourquoi Péguy se répète-t-il? Péguy est-il illisible?", republished in Camille Riquier (ed.), *Charles Péguy*, Paris: Les Cahiers du Cerf, 2014, pp. 339–63. This text was part of Latour's doctoral thesis in philosophy: *Exégèse et Philosophie*, Université de Tours, 1975.
34 Bruno Latour, *Enquête sur les Modes d'existence*, Paris: La Découverte, 2012. English translation: *An Inquiry into Modes of Existence: An Anthropology of the Moderns*, trans. Catherine Porter, Cambridge, MA: Harvard University Press, 2013; Isabelle Stengers and Bruno Latour, "The Sphinx of the Work," preface to Étienne Souriau's *The Different Modes of Existence, followed by, Of the Mode of Existence of the Work to be Made*, trans. Erik Beranek and Tim Howles, Minneapolis, MN: Univocal, 2015. This "light touch" can perhaps be associated with Deleuze and Guattari's "allusion" cited by Stengers in *Penser avec Whitehead* (p. 543) as "Philosophy as a gigantic allusion" (p. 150 of Gilles Deleuze and Félix Guattari, *Qu'est-ce que la philosophie?*, Paris: Minuit, 1991. This sentence is curiously not included on p. 159 of *What is Philosophy?* trans. H. Tomlinson and G. Burchell, New York: Columbia University Press, 1994 [Trans.]).
35 Stengers, *Activer les possibles*, p. 138.
36 The simplest definition of pragmatism is the following: "The pragmatic method is . . . to try to interpret each notion by tracing its respective practical consequences. What difference would it practically make to anyone if this notion rather than that notion were true? If no practical difference whatever can be traced, then the alternatives mean practically the same thing, and all dispute is idle." William James, *Pragmatism*, New York: Longmans, Green, 1907, p. 45.
37 See the interview with Bruno Latour by Laurent Godmer and David Smadja, "The Work of Bruno Latour: Exegetical Political Thinking," in *Raisons politiques* 47(3) (2012): 115–48, trans. JPD Systems, p. xvii. https://www.cairn-int.info/journal-raisons-politiques-2012-3-page-115.htm
38 Latour, *Politics of Nature*, p. 266.
39 Isabelle Stengers, "Entre collègues et amis," in Pierre Verstraeten and Isabelle Stengers (eds), *Gilles Deleuze*, Paris: Vrin, 1998, p. 158.
40 Stengers, *Activer les possibles*, pp. 126–7.
41 Isabelle Stengers, "Cultures: guerre et paix. Une semaine à Cerisy," *Ethnopsy. Les mondes contemporains de la guérison* 4, April 2002: 9.
42 Donna Haraway, *Staying with the Trouble: Making Kin in the Chthulucene*, Durham, NC: Duke University Press, p. 43.

Chapter 1 To De-Epistemologize . . .

1 Bruno Latour, *La Science en action*, La Découverte, Paris, 1989. English translation, *Science in Action: How to Follow Scientists and Engineers through Society*, Cambridge, MA: Harvard University Press, 1987.
2 Latour, *La Science en action*, p. 399, n. 17. (Footnote not in the English edition [Trans.].)
3 Isabelle Stengers, *L'Invention des sciences modernes*, Paris: La Découverte, 1993. English translation, *The Invention of Modern Science*, trans. Daniel Smith (Theory Out of Bounds, 19), MN: University of Minnesota Press, 2000.
4 Stengers, *The Invention of Modern Science*, p. 177, n. 10. Translation modified.
5 Carlo Ginzburg, "Morelli, Freud and Sherlock Holmes: Clues and Scientific Method," *History Workshop Journal* 9(1): 5–36, 8.
6 Ginzburg, "Morelli, Freud and Sherlock Holmes," p. 29.
7 Bruno Latour, "The Netz-Works of Greek Deductions – A Review of Reviel Netz's *The Shaping of Deduction in Greek Mathematics*," *Social Studies of Science* (2008) 38(3): 441–59.
8 Stengers, *Catastrophic Times*, p. 74. Translation modified.
9 Bruno Latour, *Un monde pluriel mais commun*, La Tour-d'Aigues: Éditions de l'Aube, 2003, p. 57.
10 See, on this point, Camille Riquier, "Charles Péguy. Métaphysique de l'événement," in Didier Debaise (ed.), *Philosophie des possessions*, pp. 197–231.
11 The first edition was with Sage in 1979, with a new edition with Princeton University Press in 1986. The French edition was with La Découverte in 1988.
12 Latour, *Rejoicing*, op. cit., and *La Fabrique du droit: Une ethnographie du conseil d'État*, Paris: La Découverte, 2002. English translation: *The Making of Law: An Ethnography of the Conseil D'État*, trans. Marina Brilman et al., Cambridge: Polity, 2010.
13 *Irreductions* is an essay included in Latour, *Pasteurization of France*, op. cit.
14 Latour, *Pasteurization of France*, p. 173.
15 Latour, *Pasteurization of France*, p. 154.
16 Latour, *Pasteurization of France*, p. 164.
17 Latour, *Pasteurization of France*, p. 202.
18 Latour, *Pasteurization of France*, p. 203.
19 Stengers, "Cultures: guerre et paix," p. 14.
20 Stengers, *Penser avec Whitehead*, pp. 25–6.
21 Latour, *Pasteurization of France*, p. 236.

22 Latour, *Pasteurization of France*, p. 161.
23 Bruno Latour, "Biography of an Inquiry: On a Book about Modes of Existence," *Social Studies of Science* 43(2): 287–301, 290.
24 Latour, *Pasteurization of France*, p. 208.
25 Latour, *Pasteurization of France*, p. 209. Translation modified.
26 Bruno Latour, "Comment redistribuer le Grand Partage?" *Revue de synthèse* 110 (April–June, 1983): 203–36.
27 Cited in Étienne Balibar, *Passions du concept. Épistémologie, théologie et politique*, *Écrits II*, Paris: La Découverte, 2020, pp. 39–40. We know about the importance that Stengers will constantly give to "common sense" in all her work, making a political point about it: "The impunity they feel in rejecting common sense makes idiots of 'those who know.'" Stengers, *Réactiver le sens commun*, p. 11. In a book co-written with Vinciane Despret, *Women Who Make a Fuss: The Unfaithful Daughters of Virginia Woolf* (Minneapolis: University of Minnesota Press, 2014), pp. 64–5, they explain how they differ from Deleuze on this question of common sense. In all of Latour's work, one can find the opposition between "good sense," which orients us to the past, and "common sense," which looks to the future.
28 If Latour began with an ethnography of a laboratory, Stengers, after she had studied chemistry, then philosophy, would become an "apprentice philosopher" (as she put it) in Ilya Prigogine's laboratory (Ilya Prigogine and Isabelle Stengers, *La Nouvelle Alliance. La métamorphose de la science*, Paris: Gallimard, 1986 [1979]. English translation, *Order Out of Chaos: Man's New Dialogue with Nature*, London: Verso, 1984).
29 "Scientists in the lab . . . know this well, but as soon as they set out to reflect on what they do, they pronounce the words that sociologists and epistemologists . . . put in their mouths." Bruno Latour, *We Have Never Been Modern*, trans. C. Porter, Cambridg, MA: Harvard University Press, 1993, p. 122. I, too, have been struck by the incapacity (at first glance, it could almost pass for a psychological block!) of researchers in the pharmaceutical industry when they were asked what they were doing. They quickly took refuge behind more or less epistemological considerations, leaning more heavily on what they had read than on their real work, as if the latter lacked dignity. This was the start of my interest in sciences as they were being done, and I was meandering a long time before meeting Stengers in 1989, who urged me to read *Laboratory Life* post-haste.
30 Isabelle Stengers, *La Vierge et le Neutrino: Les scientifiques dans la tourmente*, Les Empêcheurs de penser en rond series, Paris: Seuil, 2006, p. 26.

31 Latour, *Pasteurization of France*, p. 228. Translation modified.
32 Bruno Latour, "Le travail de l'image ou l'intelligence savante redistribuée," in *Petites Leçons de sociologie des sciences*, Paris: La Découverte, 2006, pp. 145–70.
33 In all fairness, Latour also provisionally borrows the words "theory-reified" (*théorie-réifiée*) to designate what he will later call black boxes.
34 It is curious that even the currents of Marxism most distant from Althusser will end up defending the same position as his, being happy with subtle distinctions (Marxism is not "a science," but is nonetheless "scientific" . . .) On Althusser's relationship to psychoanalysis, see Léon Chertok, Didier Gille, and Isabelle Stengers, *Une vie de combats: De l'antifascisme à l'hypnose*, Paris: La Découverte/Poche, 2020.
35 As an example, he congratulates Timothy Mitchell's *Carbon Democracy: Political Power in the Age of Oil*, London: Verso, 2013, on being a truly materialist approach.
36 Latour, *Pasteurization of France*, p. 156. Translation modified.
37 Latour, *Pasteurization of France*, p. 158.
38 Pierre Bourdieu, *Science de la science et réflexivité*, Paris: Raisons d'agir, Cours & Travaux series, 2001, p. 208. Not in Pierre Bourdieu, *Science of Science and Reflexivity*, trans. R. Nice, Chicago: University of Chicago Press, 2004.
39 *La vie de laboratoire*, p. 16. He has also written, "a Bachelardian sharpens the 'epistemological break' to guillotine those who have not yet 'found the sure Path of a science.'" Latour, *Pasteurization of France*, p. 163. Translation modified.
40 Bruno Latour, "Il ne faut pas qu'une science soit fermée ou ouverte," interview with Jean-Marc Lévy-Leblond, *Rue Descartes* 41 (2003): 66–81.
41 Latour, *Un monde pluriel mais commun*, p. 27.
42 Stengers, *Thinking with Whitehead*, p. 486.
43 The Pasteur story runs right through Latour's *oeuvre*, and this particular controversy is treated in Latour's *Pasteur: une science, un style, un siècle*, Paris: Perrin, 1994.
44 Latour, *Science in Action*, p. 184.
45 Latour, *Vie de laboratoire*, p. 218.
46 Latour, *We Have Never Been Modern*, p. 92.
47 Latour, "Comment redistribuer le Grand Partage."
48 Latour recognizes the fact that it is the historians who have paved the way more than many others. He lobbied for the translation and French publication of Ludwik Flek's *Genesis and Development of a Scientific Fact*, trans. Frederick Bradley and Thaddeus J. Trenn, Chicago: University of Chicago Press, 1979 (1935).

49 Latour, "Do Scientific Objects Have a History? Pasteur and Whitehead in a Bath of Lactic Acid," trans. Lydia Davis, in *Common Knowledge* 5(1) (Spring 1996): 76–91, 79.
50 Latour, *Pasteurization of France*, p. 216.
51 Latour, "Biography of an Inquiry," p. 294.
52 Latour, *Pasteurization of France*, p. 169.
53 Steven Shapin and Simon Schaffer, *Leviathan and the Air-Pump: Hobbes, Boyle, and the Experimental Life*, Princeton, NJ: Princeton University Press, 1985.
54 Latour, *We Have Never Been Modern*, p. 23.
55 Latour, "Do Scientific Objects Have a History?" p. 91. Translation modified.
56 Latour, "Do Scientific Objects Have a History?" pp. 89–90. Translation modified.
57 Bruno Latour, "Do Scientific Objects Have a History?" pp. 82–5.
58 Stengers, *La Vierge et le Neutrino: Les scientifiques dans la tourmente*, p. 105.
59 Stengers, *La Vierge et le Neutrino: Les scientifiques dans la tourmente*, p. 106.
60 Bruno Latour, "On Interobjectivity," in *Mind, Culture, and Activity: An International Journal* 3(4) (1996): 228–45, 246–69.
61 Bruno Latour, *On the Modern Cult of the Factish Gods*, Durham, NC: Duke University Press, 2010, p. 129, n. 17.
62 Latour, *Science in Action*, p. 70.
63 Latour, *Science in Action*, p. 72.
64 Latour, *Science in Action*, p. 73.
65 Stengers, *Invention of Modern Science*, p. 84.
66 Latour, *Pasteurization of France*, p. 185.
67 Latour, *Pasteurization of France*, pp. 170–1.
68 Latour, *After Lockdown*, p. 4. Translation modified.
69 What is, a priori, in common between the photograph of an isolated neuron taken, at the start of a contribution, during an experimental procedure on a rat brain by a lab technician, and a graph at the end of the article that shows the electrical and chemical characteristics of this neuron? It is this continuity, only separated by the hiatuses of successive reinscriptions that separate them, this "step by step" movement that the researchers must never lose sight of.
70 Latour, "Netz-Works of Greek Deductions," p. 17.
71 Latour, *Inquiry into Modes of Existence*, p. 39.
72 Latour, *Inquiry into Modes of Existence*, p. 40.
73 Latour, "A Textbook Case Revisited," p. 13.
74 Latour, "A Textbook Case Revisited," pp. 9, 12.
75 Latour, "A Textbook Case Revisited," p. 15.
76 Latour, *Un monde pluriel mais commun*, p. 26.

77 Bruno Latour, *Le Métier de chercheur. Regard d'un anthropologue*, Paris: INRA éditions, 1995, p. 67.
78 Latour, "A Textbook Case Revisited," p. 15.
79 Latour, *Science in Action*, pp. 109–10.
80 Particularly in *Politics of Nature*.
81 Bruno Latour, trans. C. Porter, *Aramis, or the Love of Technology*, Cambridge, MA: Harvard University Press, 1996 (1992).
82 Latour, *Aramis*, pp. 280–6.
83 Latour, *Inquiry into Modes of Existence*, p. 216.
84 Latour, *Inquiry into Modes of Existence*, p. 219.
85 Latour, *Inquiry into Modes of Existence*, p. 220.

Chapter 2 . . . Or Disamalgamate the Sciences

1 This text came out in the form of a little book, *La Volonté de faire science: À propos de la psychanalyse*, in multiple editions, the latest being Paris: Seuil, Les Empêcheurs de penser en rond series, 1992.
2 Isabelle Stengers, "Une politique de l'hérésie," interview in *Vacarme* 19 (Spring 2002): 4–13.
3 Stengers, *Activer les possibles*, pp. 31–2.
4 At the time, Stengers published two books in common with Prigogine, and an edited collection: *Order Out of Chaos*; *Entre le temps et l'éternité*, Paris: Fayard, 1988; Isabelle Stengers and Judith Schlanger, *Les Concepts scientifiques: Invention et pouvoir*, Paris: La Découverte, 1989; Isabelle Stengers (ed.), *D'une science à l'autre: Les concepts nomades*, Paris: Seuil, 1987.
5 Two main books came out of this partnership: Léon Chertok and Isabelle Stengers, *Le Cœur et la Raison. L'hypnose en question de Lavoisier à Lacan*, Paris: Payot, 1989; English translation by Martha Noel Evans: *A Critique of Psychoanalytic Reason*, Stanford, CA: Stanford University Press, 1992; Léon Chertok, Isabelle Stengers, and Didier Gille, *Mémoires d'un hérétique*, Paris: La Découverte, 1990 (new edition with the title, *Une vie de combats*, op. cit.).
6 See also, Léon Chertok and Isabelle Stengers, *L'Hypnose, blessure narcissique*, Paris: La Découverte, Les Empêcheurs de penser en rond series, 1990, taken from a lecture given to the URSS Academy of Sciences.
7 More literally, "the will to do science" (*la volonté de faire science*) [Trans].
8 Stengers, *La Volonté de faire science*, p. 12.
9 Stengers, *La Volonté de faire science*, p. 9.
10 Personal communication with the author.

11 Stengers, *Penser avec Whitehead*, p. 17. Not in the English translation.
12 *La Vierge et le Neutrino: Les scientifiques dans la tourmente*, p. 48.
13 Isabelle Stengers, "The Doctor and the Charlatan," in Tobie Nathan and Isabelle Stengers, *Doctors and Healers*, trans. S. Muecke, Cambridge: Polity Press, 2018, p. 120.
14 Stengers, "The Doctor and the Charlatan," p. 103.
15 Stengers, *Activer les possibles*, p. 6.
16 Isabelle Stengers, *Another Science is Possible: A Manifesto for Slow Science*, trans. S. Muecke, Cambridge: Polity Press, 2018, p. 57.
17 Isabelle Stengers, "Un engagement pour le possible," in the journal *Cosmopolitiques* 1, "La nature n'est plus ce qu'elle était," April 2006.
18 Stengers, *La Vierge et le Neutrino: Les scientifiques dans la tourmente*, p. 210.
19 Latour, *Pasteurization of France*, p. 232.
20 Bruno Latour, Laurent Godmer, and David Smadja, "L'œuvre de Bruno Latour: une pensée politique exégétique," op. cit.
21 Stengers, *La Volonté de faire science*, pp. 67–8.
22 Stengers, *La Volonté de faire science*, p. 37.
23 Stengers, *La Volonté de faire science*, p. 82.
24 Latour, *Pasteurization of France*, p. 214.
25 Stengers, *La Vierge et le Neutrino: Les scientifiques dans la tourmente*, p. 165.
26 Stengers, *Another Science*, pp. 30–1. The actual quotation from Bachelard is the following: "Science is totally opposed to opinion, not just in principle but equally in its need to come to full fruition. If it happens to justify opinion on a particular point, it is for reasons other than those that are the basis of opinion; opinion's right is therefore always to be wrong." Gaston Bachelard, *The Formation of the Scientific Mind*, trans. Mary McAllester Jones, Manchester: Clinamen Press, p. 25.
27 Stengers, *Another Science,* p. 116. Note that Isabelle Stengers wrote a history of chemistry with Bernadette Bensaude-Vincent: *Histoire de la chimie*, Paris: La Découverte, 2001.

Chapter 3 A Brief Exercise in Empirical Philosophy

1 See Bruno Latour, "Le 'pédofil' de Boa Vista – montage photo-philosophique," *La Clef de Berlin et autres leçons d'un amateur de sciences*, Paris: La Découverte, 1993, pp. 171–225, with a paperback edition with the same pagination under the title, *Petites*

Leçons de sociologie des sciences, Paris: La Découverte, 2006 (1996). The Brazilian study appears in English as chapter 2 of *Pandora's Hope*, "Circulating Reference: Sampling the Soil in the Amazon Forest," pp. 24–79 [Trans.].
2 Latour, "Circulating Reference," p. 53.
3 Latour, "Circulating Reference," p. 48.
4 Latour, "Circulating Reference," p. 49.
5 Latour, "Circulating Reference," p. 63.
6 Latour, "Circulating Reference," p. 58.
7 Latour, "Circulating Reference," p. 64.
8 Latour, "Circulating Reference," p. 56.
9 Latour, "Circulating Reference," p. 65.
10 Latour, "Circulating Reference," p. 66.
11 Latour, "Circulating Reference," pp. 78–9.
12 In *Laboratory Life*.
13 Latour, "Circulating Reference," p. 43.
14 Latour, "Circulating Reference," p. 30.
15 Latour, "Circulating Reference," pp. 25–7.
16 Latour, "Circulating Reference," p. 76.
17 Bruno Latour, "Reflections on Etienne Souriau's Les différents modes d'existence," trans. S. Muecke, in Graham Harman, Levi Bryant, and Nick Srnicek (eds), *The Speculative Turn*, Melbourne: re.press, 2011, pp. 306–7.
18 Stengers, *Thinking with Whitehead*, p. 518. Translation modified.
19 Stengers, *Catastrophic Times*, p. 29, n. 2.
20 Stengers, *Catastrophic Times*, p. 91. Translation modified.
21 Latour, "Biography of an Inquiry," p. 295. Translation modified.
22 Stengers, "Entre collègues et amis," p. 160.
23 Stengers, *La Vierge et le Neutrino: Les scientifiques dans la tourmente*, p. 173.
24 Bruno Latour, *Reassembling the Social: An Introduction to Actor-Network Theory*. Oxford: Oxford University Press, 2005.

Chapter 4 Sociology or Politics?

1 Latour, *Pasteurization of France*, p. 205.
2 Bruno Latour and Vincent Antonin Lépinay, *L'Économie science des intérêts passionnés: Introduction à l'anthropologie économique de Gabriel Tarde*, Paris: La Découverte, Les Empêcheurs de penser en rond series, 2008, p. 127. English: *The Science of Passionate Interests: An Introduction to Gabriel Tarde's Economic Anthropology*, Chicago: Prickly Paradigm Press, 2009, p. 82.
3 Latour, *Science in Action*, p. 255.

4 Latour, *Pasteurization of France*, pp. 204–5. Translation modified.
5 Latour, *Irréductions*, p. 253. Not in the English version [Trans.].
6 Bruno Latour, *Le métier de chercheur: Regard d'un anthropologie*, Versailles: Éditions Quæ, 2001, p. 68.
7 Latour, *La Science en action*, p. 399, n. 17. (Note not in the English edition [Trans.].)
8 Latour, *Science in Action*, p. 265, n.15. He is referring here to Isabelle Stengers' doctoral thesis, *États et Processus*, Brussels: ULB, 1983.
9 Latour, *Science in Action*, p. 61.
10 Latour, *Science in Action*, p. 45.
11 Latour, *La Science en action*, pp. 144–5. (Citation not in the English edition [Trans.].)
12 Latour and Woolgar, *Laboratory Life*, p. 243.
13 Latour and Woolgar, *Laboratory Life*, p. 260, n. 17.
14 Stengers, *Invention of Modern Science*, p. 3.
15 But more precisely his theory of the three worlds, *Invention of Modern Science*, pp. 42–6.
16 Stengers, *Invention of Modern Science*, p. 58.
17 Stengers, *Invention of Modern Science*, p. 25. Translation modified.
18 Stengers, *Invention of Modern Science*, p. 25.
19 Stengers, *Invention of Modern Science*, p. 101.
20 Stengers, *Thinking with Whitehead*, p. 125.
21 Stengers, *Invention of Modern Science*, p. 107.
22 Stengers, *Invention of Modern Science*, p. 108. Stengers' emphasis.
23 Stengers, *La Vierge et le Neutrino: Les scientifiques dans la tourmente*, p. 47.
24 Stengers, *La Vierge et le Neutrino: Les scientifiques dans la tourmente*, p. 46.
25 Stengers, *Another Science*, p. 143.
26 Stengers, *Catastrophic Times*, pp. 70–1. Translation modified.
27 Stengers, *Making Sense in Common*, p. 116.
28 Stengers, *Making Sense in Common*, pp. 119–20.
29 Stengers, "Le laboratoire de l'ethnopsychiatrie," p. 29.
30 Stengers, *Invention of Modern Science*, p. 57.
31 Stengers, *Invention of Modern Science*, p. 61. Translation modified.
32 Stengers, *Invention of Modern Science*, p. 61.
33 Stengers, *Invention of Modern Science*, p. 64.
34 Stengers, *Invention of Modern Science*, p. 64.
35 Latour, *Science in Action*, p. 78.
36 Latour, *La Science en Action*, p. 122.
37 Stengers, *Invention of Modern Science*, p. 89.
38 Stengers, *Thinking with Whitehead*, p. 260. Translation modified.
39 Latour, *Pasteur. Une science, un style, un siècle*, p. 104.

40 Stengers, *Invention of Modern Science*, p. 64.
41 Stengers, *Invention of Modern Science*, p. 65. Translation modified.
42 Stengers, *Invention of Modern Science*, p. 65.
43 Stengers, *Invention of Modern Science*, p. 96.
44 Stengers, *Cosmopolitics II*, p. 226.
45 Stengers, *Cosmopolitics II*, p. 57.
46 Stengers, *Another Science*, pp. 53–4.
47 Donna Haraway, "Situated Knowledges: The Science Question in Feminism and the Privilege of Partial Perspective," *Feminist Studies* 14(3) (1988): 575–99.
48 Stengers, *Another Science*, p. 147.
49 Stengers, *Another Science*, p. 118.
50 Stengers, *La Vierge et le Neutrino: Les scientifiques dans la tourmente*, p. 115.
51 Isabelle Stengers, "Un engagement pour le possible," *Cosmopolitique* 1 (April 2006).
52 Stengers, *In Catastrophic Times*, pp. 35–41.

Chapter 5 The *Factish* Gods

1 Stengers, *Cosmopolitics II*, p. 228.
2 Stengers, *The Invention of Modern Science*, p. 136.
3 Stengers, *The Invention of Modern Science*, p. 136.
4 Stengers, *The Invention of Modern Science*, p. 138. Translation modified.
5 Stengers, *The Invention of Modern Science*, p. 138.
6 Stengers, *Cosmopolitics II*, p. 231.
7 Stengers, *Cosmopolitics II*, p. 229.
8 Stengers, *The Invention of Modern Science*, p. 139.
9 Stengers, *Thinking with Whitehead*, p. 129. Translation modified.
10 Stengers, *The Invention of Modern Science*, p. 140. Translation modified.
11 Carlo Ginzburg, *Clues, Myths, and the Historical Method*, trans. John and Anne C. Tedeschi, Baltimore: Johns Hopkins University Press, 1989, p. 102.
12 Stengers, *The Invention of Modern Science*, p. 143.
13 Stengers, *Another Science*, p. 18.
14 Carla Hustak and Natasha Myers, "Involutionary Momentum: Affective Ecologies and the Sciences of Plant/Insect Encounters," *differences: A Journal of Feminist Cultural Studies* 23(3) (2012): 106.
15 See, in particular, *The Dance of the Arabian Babbler: Birth of an Ethological Theory*, trans. J. Bussolini, Minneapolis: Univocal, 2021;

and *Quand le loup habitera avec l'agneau,* Paris: La Découverte, Les Empêcheurs de penser en rond series, 2020.
16 Stengers, *The Invention of Modern Science*, pp. 148, 167.
17 Stengers, *The Invention of Modern Science*, p. 147. Stengers' emphasis.
18 Stengers, *Cosmopolitics II*, p. 214.
19 Bruno Latour, "Des sujets récalcitrants," in *Chroniques d'un amateur de sciences*, Paris: Presses des Mines, 2006.
20 Stengers, *Another Science*, pp. 65–6. Translation modified.
21 Claude Lévi-Strauss, "The Effectiveness of Symbols," in *Structural Anthropology*, New York: Doubleday, 1949.
22 Isabelle Stengers, "Le laboratoire de l'ethnopsychiatrie." Preface to Tobie Nathan, *Nous ne sommes pas seuls au monde*, Paris: Seuil, Les Empêcheurs de penser en rond series, 2001, pp. 38, 31.
23 Stengers, *Cosmopolitics II*, p. 327.
24 Stengers, *Cosmopolitics II*, p. 326. She is citing from Tobie Nathan, *L'influence qui guérit*, Paris: Odile Jacob, 1994, p. 18.
25 Nathan and Stengers, *Doctors and Healers*, p. 84.
26 Stengers, "Le laboratoire de l'ethnopsychiatrie," pp. 20–1.
27 Stengers, "The Doctor and the Charlatan," in Nathan and Stengers, *Doctors and Healers*, pp. 122–3. Her reference to the "Royal Way" is to Jacques Derrida, "Plato's Pharmacy," in *Dissemination*, trans. B. Johnson, Chicago: Chicago University Press, 1981.
28 Stengers, *L'Hypnose entre science et magie*, p. 58.
29 Stengers, "The Doctor and the Charlatan," p. 126.
30 Tobie Nathan, "Manifeste pour une psychothérapie démocratique," in Tobie Nathan (ed.), *La Guerre des psys: Manifeste pour une psychothérapie démocratique*, Paris: Seuil, Les Empêcheurs de penser en rond series, 2006, p. 23.
31 Stengers, *L'Hypnose entre science et magie*, p. 160.
32 Stengers, *Making Sense in Common*, p. 165.
33 Stengers wrote: "What the patients consent to is an adventure, a bet on a becoming or on a possible metamorphosis." Stengers, "Le laboratoire de l'ethnopsychiatrie," preface to Tobie Nathan, *Nous ne sommes pas seuls au monde*, Paris: Seuil, Les Empêcheurs de penser en rond, 2001, p. 35. In the *Inquiry into the Modes of Existence*, Latour will make metamorphoses (MET) into a separate mode of existence. But here he is speaking, in a provisional manner, of "transfears" [*trans-frayeurs*], the word *frayeur* being directly borrowed from Nathan. See Latour, *Factish Gods*.
34 Bruno Latour, *Factish Gods*, p. 136.
35 Latour, *Factish Gods*, p. 48. What followed was the creation of a collective composed of Isabelle Stengers, Bruno Latour, Tobie Nathan, Bruno Pinchard, and myself, a monthly seminar following

on from the one that Isabelle Stengers was holding for years with Léon Chertok, devoted to hypnosis.
36 Latour, *Factish Gods*, p. 66.
37 Latour, *Factish Gods*, p. 35.
38 This is how the work of Marie-Rose Moro could be characterized.
39 Latour, *Pasteurization of France*, pp. 187–8. Translation modified.
40 Latour, *Politics of Nature*, p. 284, n. 40. Translation modified. The 1994 reference to Nathan is to *L'influence qui guérit*.
41 Didier Fassin, "L'ethnopsychiatrie et ses réseaux. L'influence qui grandit," *Genèses* 35, 1999. See Tobie Nathan's reply, "Psychothérapie et politique. Les enjeux théoriques, institutionnels et politiques de l'ethnopsychiatrie," in his *Nous ne sommes pas seuls au monde*, pp. 69–110.
42 Bruno Latour and Isabelle Stengers, "Du bon usage de l'ethnopsychiatrie." *Libération*, January 21, 1997.
43 Latour, *Rejoicing*, pp. 155–6.
44 Bruno Latour, "Factures/Fractures: From the Concept of Network to the Concept of Attachment," *RES: Anthropology and Aesthetics* 36 (Autumn 1999), p. 22. Translation modified.
45 Latour, *Factish Gods*, p. 12.
46 Latour, *Factish Gods*, p. 131, n. 26. In explicit reference to Pierre Bourdieu.
47 Stengers, *Catastrophic Times*, pp. 112–13. Translation modified.
48 Latour, *Pasteurization of France*, pp. 178–9.
49 Latour, *Sur le culte moderne des dieux faitiches*, p. 58. Not in English edition.
50 Latour, *Politics of Nature*, p. 225.
51 Stengers, *Catastrophic Times*, p. 34.
52 Philippe Pignarre and Isabelle Stengers, *Capitalist Sorcery: Breaking the Spell*, trans. and ed. A. Goffey, Basingstoke: Palgrave Macmillan, 2011, p. 106. Translation modified. Philippe Pignarre notes: in order to make things easier, this book will henceforth be cited under the name of Stengers alone because the chosen citations will be more typically Stengersian.
53 *Factish Gods*, p. 18.
54 *Factish Gods*, p. 33.
55 Medicines, as a very general kind of technical object, are perfect examples of "tricks."
56 Latour, *Factish Gods*, p. 135, n. 51.
57 Pignarre and Stengers, *Capitalist Sorcery*, pp. 137–8.
58 Latour, *Factish Gods*, pp. 41–2.
59 Latour, *Factish Gods*, p. 136, n. 53. Translation modified.
60 Stengers, *Cosmopolitics II*, p. 371.
61 Latour, *Factish Gods*, p. 129, n. 17.

62 Latour, *Factish Gods*, p. 64. Translation modified.
63 Latour, "Biography of an Inquiry," p. 292.
64 Latour, "Biography of an Inquiry," p. 297.
65 I am citing them here only to whet the appetite of the reader to pick up the work in question: *Cosmopolitics II*, pp. 62, 371, 383, 414.
66 Kyle Harper, *The Fate of Rome: Climate, Disease and the End of an Empire*, Princeton, NJ: Princeton University Press, 2017.
67 Latour, *Pasteurization of France*, p. 35. Translation modified.
68 Latour, *Science in Action*, p. 143.
69 Latour, *Reassembling the Social*, pp. 30–1.
70 Latour, *Modes of Existence*, p. 296.
71 She writes on the last page of her *Invention of Modern Science*, "As for myself, I will explain this perspective [the parliament of things] one day in terms derived from the work of Alfred North Whitehead," p. 178, n. 18.
72 Stengers, *L'Hypnose entre science et magie*, p. 148, n. 83.
73 Stengers, *L'Hypnose entre science et magie*, pp. 109–10, n. 53.
74 Stengers, *L'Hypnose entre science et magie*, pp. 109–10.
75 Stengers, *Thinking with Whitehead*, p. 136.
76 Stengers, *L'Hypnose entre science et magie*, pp. 134–5.
77 Stengers, *L'Hypnose entre science et magie*, p. 140.
78 Stengers, *L'Hypnose entre science et magie*, p. 141.
79 Stengers, *L'Hypnose entre science et magie*, p. 135.
80 Tobie Nathan, *Du Commerce avec les diables*, Paris: Seuil, Les Empêcheurs de penser en rond series, 2004.
81 Stengers, *L'Hypnose entre science et magie*, p. 136.
82 Stengers, *L'Hypnose entre science et magie*, p. 14.
83 Stengers, *L'Hypnose entre science et magie*, p. 14.
84 Stengers, *L'Hypnose entre science et magie*, p. 151.
85 Stengers, *L'Hypnose entre science et magie*, p. 151.
86 Stengers, *L'Hypnose entre science et magie*, pp. 152–4.
87 Nathan, Stengers, and Hounkpatin, *La Damnation de Freud*, Act IV, Sc. 4, p. 134. They wrote this play together, imagining an encounter between Freud and an apprentice African healer (Ekudi) at the end of World War I. Ekudi calls Freud "Fofo."

Chapter 6 The Parliament of Things: Doing Ecology

1 Latour, *Politics of Nature*, p. 1.
2 Latour, *Politics of Nature*, pp. 227–8.
3 Shapin and Schaffer, *Leviathan and the Air-Pump*.
4 Latour, *We Have Never Been Modern*, p. 23.

5 Latour, *We Have Never Been Modern*, p. 29.
6 Cited in Latour, *Pasteur: Une science, un style, un siècle*, p. 166.
7 Latour, *We Have Never Been Modern*, p. 71.
8 Latour, *We Have Never Been Modern*, p. 97.
9 Latour, *We Have Never Been Modern*, pp. 101, 103.
10 Latour, *We Have Never Been Modern*, p. 108.
11 Latour, *We Have Never Been Modern*, pp. 113–14.
12 Latour, *We Have Never Been Modern*, p. 135.
13 Stengers, *Another Science*, p. 79.
14 Stengers, *Invention of Modern Science*, pp. 127–8.
15 Stengers, *Invention of Modern Science*, p. 128. Translation modified.
16 Stengers, *Another Science*, p. 55.
17 Bruno Latour, "Différencier amis et ennemis à l'époque de l'Anthropocène," in Debaise and Stengers (eds), *Gestes spéculatifs*.
18 Latour, *Politics of Nature*, pp. 105–6.
19 Latour, *Politics of Nature*, p. 250.
20 Latour, *We Have Never Been Modern*, p. 143.
21 This is why Latour could not join the project of a citizens' convention on the climate, the aim of which was precisely to obtain, by way of a random draw of the participants, "naked" citizens. Stengers upholds a somewhat different point of view with the suggestion of "generative apparatuses" that can allow "new possibilities for speaking and feeling to emerge. It makes possible the transformation of antagonistic reasons into contrasts that matter." The random draw could be a precious part of such an apparatus as it turns the implied nakedness into a garment which puts demands on the wearer. It is to be noted that a generative apparatus is without guarantees: it is a pharmakon, one Macron might have had the bitter experience of. Stengers, *Making Sense in Common*, p. 61.
22 Bruno Latour, "Esquisse d'un parlement des choses," *Écologie et Politique* 56 (2018): 47–64.
23 Latour, "Différencier amis et ennemis à l'époque de l'Anthropocène," p. 30.
24 Latour, "Différencier amis et ennemis à l'époque de l'Anthropocène," pp. 30–1.
25 Latour, "Différencier amis et ennemis à l'époque de l'Anthropocène," p. 31.
26 Bruno Latour, "To Modernize or to Ecologize? That's the Question," in N. Castree and B. Willems-Braun (eds), *Remaking Reality: Nature at the Millennium*, London and New York: Routledge, 1998, p. 230. Latour happily acknowledges this new engagement: "The seeming paradox in the fact that the so-called question of the environment appeared only when the external environment disappeared was what led me to investigate these ecological questions,

in the context of a study of the implementation of a new law on water in France." Bruno Latour, *Facing Gaia: Eight Lectures on the New Climatic Regime*, trans. C. Porter, Cambridge: Polity, p. 36, n. 69.
27 Stengers, *Cosmopolitics II*, p. 394. Translation modified.
28 Stengers, Un engagement pour le possible," p. 30.
29 Stengers, *Cosmopolitics II*, p. 394.
30 Latour, *Politics of Nature*, pp. 155–6.
31 Stengers, "Cultures: guerre et paix," p. 29.
32 Stengers, "Cultures: guerre et paix," p. 28.
33 Stengers, "Un engagement pour le possible," p. 30.
34 Stengers, "Une politique de l'hérésie."
35 Stengers, *Cosmopolitics II*, p. 355.
36 Latour, *Pasteurization of France*, p. 221. Translation modified.
37 Stengers, *Cosmopolitics II*, p. 355.
38 Stengers, *Cosmopolitics II*, p. 356. Translation modified.
39 Stengers, *Cosmopolitics II*, p. 362.
40 Latour, *Rejoicing*, p. 178.
41 Stengers, *Cosmopolitics II*, p. 375.
42 Stengers, "Cultures: guerre et paix," p. 20.
43 Stengers, *La Vierge et le Neutrino: Les scientifiques dans la tourmente*, p. 247.
44 Stengers, *Penser avec Whitehead*, p. 543.
45 Stengers, *Thinking with Whitehead*, p. 496.
46 Latour, *Politics of Nature*, p. 164.
47 Isabelle Stengers, "L'insistance du possible," in Didier Debaise and Isabelle Stengers (eds), *Gestes spéculatifs*, pp. 5–22. See a related article in English, Isabelle Stengers, "The Insistence of Possibles; Towards a Speculative Pragmatism," in *Parse Journal* 7 (Autumn) 2017 [Trans.].
48 Stengers, *La Vierge et le Neutrino: Les scientifiques dans la tourmente*, p. 252.
49 Stengers, *Cosmopolitics II*, p. 382.
50 Personal communication with the author.
51 Stengers writes: "It is appropriate here to distinguish 'influence,' in Nathan's sense, from 'suggestion,' 'imagination,' or 'placebo effect,' because these three terms – even if they are not defined pejoratively in line with the pervasive theme of the irrational – designate an ingredient held to be 'natural,' 'psychological,' 'found everywhere,' and not a technical thought likely to bring specific teaching to the art of curing." *Doctors and Healers*, pp. 122–3.
52 These series of questions relate to the two chambers that constitute the Parliament of Things in *Politics of Nature*, p. 167.
53 Stengers, "Cultures: guerre et paix," pp. 20–3.

54 Stengers, *La Vierge et le Neutrino: Les scientifiques dans la tourmente*, p. 126.
55 Stengers, *La Vierge et le Neutrino: Les scientifiques dans la tourmente*, p. 127.
56 This is the kind of inquiry she will carry out with Vinciane Despret in *Women Who Make a Fuss*, pp. 85–7.
57 Stengers, *Thinking with Whitehead*, p. 484.
58 Isabelle Stengers, "Philosophie activiste, récits spéculatifs et ouverture des possibles," interview with Véronique Bergen, *Le Carnet et les Instants. Le blog des lettres belges francophones*, n.d., online.
59 Stengers makes clear her disagreement with Deleuze on this matter of "common sense," which she derives from Leibniz.
60 Stengers, *Thinking with Whitehead*, p. 420.
61 Stengers, "Introduction," in Isabelle Stengers (ed.), *L'Effet Whitehead*, Paris: Vrin, 1994, p. 18.
62 Latour, *Politics of Nature*, p. 242. Translation modified.
63 Stengers, "Un engagement pour le possible," p. 30.
64 Stengers, *Catastrophic Times*, pp. 18–22.
65 Stengers, *Invention of Modern Science*, p. 153.
66 Stengers, *Invention of Modern Science*, pp. 153–4. "The concept is the contour, the configuration, the constellation of an event to come. Concepts in this sense belong to philosophy by right, because it is philosophy that creates them and never stops creating them." Deleuze and Guattari, *What is Philosophy?* pp. 32–3.
67 Stengers, *Invention of Modern Science*, p. 159.
68 Stengers, *Invention of Modern Science*, p. 163.
69 Stengers, *Cosmopolitics I*, pp. vii–viii.
70 This famous statement from Cromwell was later explained further by Stengers: "Even if it was addressed to the Church of Scotland [in 1650], which had pledged allegiance to the royalist cause, by Cromwell, as the head of an army about to defeat the Scottish, this call has indeed echoed as if carrying no such memory, its efficacy fully in the now of each of its moments of reception." Isabelle Stengers, "Awakening the Call of Others: What I Learned from Existential Ecology," in Thom van Dooren and Matthew Chrulew (eds), *Kin: Thinking with Deborah Bird Rose*, Durham: Duke University Press, 2022, p. 66 [Trans.].
71 Stengers, "Une politique de l'hérésie."
72 Bruno Latour, "Whose Cosmos, Which Cosmopolitics? Comments on the Peace Terms of Ulrich Beck," *Common Knowledge* 10(3) (2004): 450–62.
73 Latour, "Whose Cosmos, Which Cosmopolitics?' Translation modified.

74 Bruno Latour, *Cogitamus: Six lettres sur les humanités scientifiques*, Paris: La Découverte, 2010, p. 121.
75 Latour, *Cogitamus*, pp. 168–9.
76 Latour, *Cogitamus*, p. 171.

Chapter 7 Identifying Modes of Existence, Thinking with Whitehead

1 Latour, *The Making of Law*.
2 Latour, *The Making of Law*, p. 129.
3 Latour, *The Making of Law*, p. 276. Translation modified.
4 Latour, "Biography of an Inquiry," p. 289.
5 Latour, *Rejoicing*, p. 180.
6 Latour, "Biography of an Inquiry," p. 289.
7 Latour, *Rejoicing*, p. 22.
8 In "Biography of an Inquiry," he summarizes his trajectory through all the inquiries carried out in the domains of law, religion, psychism, about which we have spoken.
9 Latour, *Inquiry into Modes of Existence*, p. 477.
10 Étienne Souriau, *The Different Modes of Existence, followed by Of the Mode of Existence of the Work-to-be-Made*, trans. E. Beranek and T. Howles, Minnesota: Univocal, 2015, with the Preface, "The Sphinx of the Work" by Isabelle Stengers and Bruno Latour.
11 Stengers, *Thinking with Whitehead*, p. 242.
12 Bruno Latour, "What is Given in Experience? A Review of Isabelle Stengers, *Penser avec Whitehead*," *Boundary* 32(1) (Spring 2005): 222–3. Translation modified.
13 Stengers, *Penser avec Whitehead*, pp. 20–1. This passage is not in the English edition.
14 Stengers, *Making Sense in Common*, p. 35. Whitehead speaks of it thus: "Another way of phrasing this theory which I am arguing against is to bifurcate nature into two divisions, namely into the nature apprehended in awareness and the nature which is the cause of awareness. The nature which is the fact apprehended in awareness holds within it the greenness of the trees, the song of the birds, the warmth of the sun, the hardness of the chairs, and the feel of the velvet. The nature which is the cause of awareness is the conjectured system of molecules and electrons which so affects the mind as to produce the awareness of apparent nature. The meeting point of these two natures is the mind, the causal nature being influent and the apparent nature being effluent." Alfred North Whitehead, *The Concept of Nature*, Project Gutenberg Ebook, 2016 [1919], p. 31.
15 Stengers, *Penser avec Whitehead*, p. 23. Didier Debaise has a

marvellous little book on the bifurcation of nature: *L'Appât des possibles: Reprise de Whitehead*, Dijon: Les Presses du réel, 2015. See also his *Un Empirisme spéculatif: Lecture de Procès et Réalité de Whitehead*, Paris: Vrin, 2006.
16 Stengers, *Penser avec Whitehead*, p. 21.
17 Stengers, *Thinking with Whitehead*, p. 40.
18 Stengers, *Making Sense in Common*, p. 37.
19 Stengers, "L'insistance du possible," p. 12.
20 Stengers, *Making Sense in Common*, p. 53.
21 Stengers, *Making Sense in Common*, p. 45.
22 Stengers, *Making Sense in Common*, p. 51.
23 Stengers, *Making Sense in Common*, p. 90.
24 Alfred North Whitehead, *Nature and Life*, Cambridge: Cambridge University Press, 2011 (1934), p. 66.
25 Stengers, *Making Sense in Common*, p. 71.
26 Stengers, *Making Sense in Common*, p. 27.
27 Stengers, *Making Sense in Common*, p. 22.
28 Stengers, *Making Sense in Common*, p. 165.
29 Latour, *Pandora's Hope*, p. 14.
30 Stengers, *Thinking with Whitehead*, p. 486.
31 Stengers, *Thinking with Whitehead*, p. 107.
32 Stengers, *Thinking with Whitehead*, p. 136.
33 Stengers, *Thinking with Whitehead*, p. 516.
34 This was during a seminar preparing the collective edition, Michel Serres (ed.), *Éléments d'histoire des sciences*, Paris: Bordas, 1993.
35 Latour, "Biography of an Inquiry," pp. 297–8.
36 Latour, "Biography of an Inquiry," pp. 297–8.
37 Stengers and Latour, "The Sphinx of the Work," p. 72.
38 Stengers and Latour, "The Sphinx of the Work," p. 21. Translation modified. Stengers and Latour cite Souriau: "The conquest of our thought goes hand in hand with that of the external world; they are both one and the same operation," p. 24.
39 Stengers and Latour, "The Sphinx of the Work," p. 21.
40 Stengers and Latour, "The Sphinx of the Work," p. 22.
41 Latour, *Factish Gods*, p. 123.
42 Stengers and Latour, "The Sphinx of the Work," p. 57.
43 Didier Debaise writes, "An inheritance of Whitehead's idea that persistence involves a trajectory of reprises can be found in Bruno Latour's *Inquiry into Modes of Existence*." *Nature as Event: The Lure of the Possible*, trans. M. Halewood, Durham, NC: Duke University Press. 2017, p. 91, n. 82.
44 Latour, *Inquiry into Modes of Existence*, p. 40.
45 Stengers and Latour, "The Sphinx of the Work," p. 44.
46 Stengers and Latour, "The Sphinx of the Work," p. 48.

47 Stengers and Latour, "The Sphinx of the Work," p. 49.
48 Latour, *Inquiry into Modes of Existence*, p. 56.
49 Stengers and Latour, "The Sphinx of the Work," p. 53.
50 Latour, *Inquiry into Modes of Existence*, p. 56.
51 Latour, *Inquiry into Modes of Existence*, p. 370.
52 Latour, *Inquiry into Modes of Existence*, p. 488.
53 Stengers and Latour, "The Sphinx of the Work," p. 36.
54 Stengers and Latour, "The Sphinx of the Work," p. 40.
55 Stengers and Latour, "The Sphinx of the Work," p. 45.
56 On the question of soul, see David Lapoujade, "Étienne Souriau. Une philosophie des existences moindres," in Didier Debaise (ed.), *Philosophie des possessions*, pp. 167–96.
57 Stengers and Latour, "The Sphinx of the Work," p. 58.
58 Stengers and Latour, "The Sphinx of the Work," p. 61.
59 Latour, *Inquiry into Modes of Existence*, p. 366.
60 Latour, *Inquiry into Modes of Existence*, p. 7.
61 Cited in Stengers, *Thinking with Whitehead*, p. 485, from Gilles Deleuze and Félix Guattari, *What is Philosophy?*, trans. H. Tomlinson and G. Burchell, New York: Columbia University Press, 1994, p. 74. Translation modified.
62 Stengers, *Making Sense in Common*, p. 109.
63 Stengers, "Que vas-tu faire de moi?" p. 64.
64 Stengers, "Que vas-tu faire de moi?" p. 66.
65 Stengers, *Making Sense in Common*, pp. 103–7.
66 Stengers, "Que vas-tu faire de moi?" p. 83.
67 Stengers, "Que vas-tu faire de moi?" p. 83.
68 Latour, *Inquiry into Modes of Existence*, p. 162. Translation modified.
69 Stengers, "Que vas-tu faire de moi?" p. 84.
70 Stengers, "Que vas-tu faire de moi?" p. 76.
71 Stengers, "Que vas-tu faire de moi?" p. 70. (This is "betrayal" in the "*traduttore, traditore*" sense [Trans.].)
72 If he has to be betrayed, it is "going by way of what made him great," she writes on p. 76. Houria Bouteldja should definitely be cited here to bring up the question of the "I" broached in the introduction to this book, and which keeps coming back: "Who is hiding behind the Cartesian 'I'? At the time the expression was formulated, America had been 'discovered' for two hundred years. Descartes is in Amsterdam, the new center of the world order. Is it conceivable to extract this 'I' from the political context of its utterance? . . . This 'I' is a conquering 'I.' It is armed. It has a gun in one hand and a Bible in the other. It is predatory. It is drunk on its victories. 'We have to become masters and possessors of nature, Descartes goes on to say.' The Cartesian 'I' asserts itself. It wants to

defy death. It will henceforth occupy the center. I think therefore I am the one who decides, who pillages, who steals, who rapes, who carries out genocide. I think therefore I am modern man, virile, capitalist, imperialist. The Cartesian 'I' will forge the philosophical foundations of whiteness." Houria Boutedja, *Whites, Jews, and Us . . . Toward a Politic of Revolutionary Love*, Semiotext, 2017, pp. 33–4. And it is indeed a man we are talking about, not a woman . . .

73 Stengers, "Que vas-tu faire de moi?," pp. 84–5.
74 Stengers, *Making Sense in Common*, p. 108.

Chapter 8 The Intrusion of Gaia

1 Appearing as ch. 6 of *Another Science is Possible*.
2 Stengers, *Catastrophic Times*, p. 47.
3 Stengers, *Thinking with Whitehead*, pp. 163–4.
4 Latour, *Inquiry into Modes of Existence*, p. xxvii.
5 Latour, *Inquiry into Modes of Existence*, p. 9.
6 Latour, "Biography of an Inquiry," p. 294. Translation modified.
7 Donna Haraway makes specific this relationship with the IPCC: "Latour *aligns* himself with the reports of the Intergovernmental Panel on Climate Change (IPCC); he does not *believe* its assessments and reports . . . He casts his lot with some worlds and worlding and not others." *Staying with the Trouble*, p. 41.
8 Latour, *Facing Gaia*, p. 33, n. 63.
9 Latour, *Pasteurization of France*, pp. 193–4. Translation modified.
10 "If . . . evolution is a 'rolling outwards,' a kind of speciation through divergence in the shape of branching trees, we approach involution as the 'rolling, curling, turning inwards' that brings distinct species together to invent new ways of life." Hustak and Myers, "Involutionary Momentum," p. 96, quoting the *Oxford English Dictionary*.
11 For the sometimes furious reactions of Anglo-Saxon sociologists of science toward Latour, Gerard de Vries's *Bruno Latour* is a useful source.
12 Félix Guattari, *The Three Ecologies*, trans. I. Pindar and P. Sutton, London: Bloomsbury, 2014.
13 Stengers, "Une politique de l'hérésie."
14 Stengers, *Thinking with Whitehead*, p. 275.
15 Stengers, *Thinking with Whitehead*, p. 496.
16 Latour, *Politics of Nature*, pp. 191–2.
17 Isabelle Stengers, *Résister au désastre*, Marseille: Wildproject, 2019, p. 64.

18 Stengers, *La Vierge et le Neutrino: Les scientifiques dans la tourmente*, p. 232, n. 1.
19 Stengers, *Résister au désastre*, p. 44.
20 Stengers, *La Vierge et le Neutrino: Les scientifiques dans la tourmente*, p. 21.
21 Stengers, *La Vierge et le Neutrino: Les scientifiques dans la tourmente*, p. 56.
22 Stengers, *La Vierge et le Neutrino: Les scientifiques dans la tourmente*, p. 69.
23 Stengers, *Une autre science est possible!*, p. 85.
24 Stengers, *La Vierge et le Neutrino: Les scientifiques dans la tourmente*, p. 98.
25 Stengers, *La Vierge et le Neutrino: Les scientifiques dans la tourmente*, p. 231.
26 Stengers, *Résister au désastre*, p. 64.
27 Stengers, *La Vierge et le Neutrino: Les scientifiques dans la tourmente*, p. 116.
28 Stengers, *Résister au désastre*, p. 56.
29 Stengers, *Making Sense in Common*, p. 99.
30 Stengers, *Making Sense in Common*, p. 98.
31 Latour, *Facing Gaia*, p. 31.
32 Bruno Latour, "Troubles dans l'engendrement," Bruno Latour interviewed by Carolina Miranda. *Revue du Crieur* No. 14, La Découverte/Mediapart, 2019, p. 64.
33 Stengers, *Another Science*, p. 39.
34 Stengers, *Making Sense in Common*, pp. 17–48.
35 Stengers, *Making Sense in Common*, p. 48.
36 Latour, *Pandora's Hope*, p. 14.
37 Latour, *Facing Gaia*, pp. 30–2.
38 Latour, *Face à Gaia*, p. 59.
39 Latour, *Facing Gaia*, p. 253, including phrases from *Face à Gaïa*, pp. 326–7.
40 Bruno Latour, "Différencier amis et ennemis à l'époque de l'Anthropocène," p. 35.
41 Patrice Maniglier, "Latour, chef de guerre: Petit traité de Gaïapolitique," in Frédérique Aït-Touati and Emanuele Coccia (eds), *Le Cri de Gaïa: Penser la Terre avec Bruno Latour*, Paris: La Découverte, Les Empêcheurs de penser en rond series, 2021, pp. 183–216.
42 Stengers, *Catastrophic Times*, pp. 48–9. Translation modified.
43 Stengers, *Catastrophic Times*, p. 51. Translation modified.
44 Stengers, *Catastrophic Times*, p. 57.
45 Latour, *Facing Gaia*, p. 194.
46 Stengers, *Thinking with Whitehead*, pp. 136–7.

47 Stengers, "Un engagement pour le possible," p. 31.
48 Stengers, "Un engagement pour le possible," p. 31.
49 Stengers, "Une politique de l'hérésie."
50 Latour, "Facture/Fractures," p. 23. Translation modified.
51 Latour, *Facing Gaia*, p. 199, n. 40.
52 Stengers, *Activer les possibles*, p. 124.
53 Stengers, *Catastrophic Times*, p. 104. Translation modified.
54 Latour, *Facing Gaia*, p. 208.
55 Latour, *Facing Gaia*, p. 209.
56 Latour, *Facing Gaia*, p. 219.
57 Karl Marx and Friedrich Engels, *The Communist Manifesto* (1888), Project Gutenberg Ebook, p. 43.
58 Latour, *Facing Gaia*, p. 198–200.
59 Latour, *Facing Gaia*, p. 198. See Eric Vogelin, *The New Science of Politics: An Introduction*, Chicago: University of Chicago Press, 1952.
60 Latour, *We Have Never Been Modern*, p. 125.
61 Latour, "On Interobjectivity," p. 240.
62 Stengers, "Un engagement pour le possible."
63 Pignarre and Stengers, *Capitalist Sorcery*, p. 13.
64 Stengers, *Catastrophic Times*, p. 53. Translation modified.
65 Pignarre and Stengers, *Capitalist Sorcery*, pp. 8–9.
66 Pignarre and Stengers, *Capitalist Sorcery*, p. 7.
67 Pignarre and Stengers, *Capitalist Sorcery*, p. 89.
68 Stengers, *Activer les possibles*, p. 45.
69 Pignarre and Stengers, *Capitalist Sorcery*, p. 140.
70 Gilles Deleuze and Félix Guattari, trans. B. Massumi, *A Thousand Plateaus; Capitalism and Schizophrenia*, Minneapolis: University of Minnesota Press, 1987, p. 377.
71 Pignarre and Stengers, *Capitalist Sorcery*, p. 123.
72 Isabelle Stengers, "La résurgence des communs." 29 April 2019, see the site: dijoncter. info.
73 Anna Lowenhaupt Tsing, *Friction, An Ethnography of Global Connection*, Princeton, NJ: Princeton University Press, 2004.
74 Pignarre and Stengers, *Capitalist Sorcery*, p. 125. Translation modified.
75 *Ravaudeurs*, which could also be translated as "scavengers" [Trans].
76 Pignarre and Stengers, *Capitalist Sorcery*, p. 136.
77 Pignarre and Stengers, *Capitalist Sorcery*, p. 137. Translation modified
78 Pignarre and Stengers, *Capitalist Sorcery*, p. 141. Translation modified.
79 Latour, *After Lockdown*, pp. 71–3.
80 Latour, *After Lockdown*, pp. 79–80. Translation modified.

81 Michel Callon, *Markets in the Making: Rethinking Competition, Goods, and Innovation*, trans. O. Custer, Princeton, NJ: Princeton University Press, 2021.
82 Latour, *Politics of Nature*, pp. 135–6.
83 Latour, *Politics of Nature*, p. 153.
84 Latour, *After Lockdown*, pp. 62–3.
85 Latour, *After Lockdown*, p. 61.
86 Latour and Lépinay, *Science of Passionate Interests*, pp. 81–2. For a more general account of Latour's relationship with Tarde, see Bruno Latour, "Gabriel Tarde. La société comme possession. La 'preuve par l'orchestre,'" in Didier Debaise (ed.), *Philosophie des possessions*, pp. 9–34.
87 Latour, *After Lockdown*, p. 64. Clearly, he is alluding here to the *Capitalist Sorcery* book.
88 Stengers, *Thinking with Whitehead*, pp. 153ff.
89 Isabelle Stengers, "Pourquoi le paysan argentin a raison de dire que le soja OGM est 'méchant,'" *Terrestres* 5, May 6, 2019.
90 Bruno Latour, "Le Covid comme crash-test," interview with Thibaut Sardier, *Libération*, May 13, 2020, available on the bruno-latour.fr site.
91 Latour, "Issues with Engendering," p. 5.
92 Latour, "Issues with Engendering," pp. 9–10.
93 Latour, *Pasteurization of France*, p. 191. Translation modified.
94 Latour, *After Lockdown*, pp. 115–17.

Conclusion: Composing a Common World . . . During the Meltdown

1 Stengers, *Thinking with Whitehead*, p. 244.
2 Stengers, *Penser avec Whitehead*, p. 544. Not in the English edition [Trans.].
3 Stengers and Despret, *Women Who Make a Fuss*, p. 105.
4 Stengers, *Thinking with Whitehead*, pp. 177–8.
5 Stengers, *Catastrophic Times*, p. 148.
6 Stengers, *Thinking with Whitehead*, p. 517.
7 Stengers, *Activer les possibles*, pp. 5–6.
8 Latour, Godmer, and Smadja, "The Work of Bruno Latour," p. xvii.
9 Latour, "Reflections on Etienne Souriau's *Les différents modes d'existence*," p. 304.
10 Stengers, *Thinking with Whitehead*, p. 519. Translation modified.
11 Stengers, *Making Sense in Common*, p. 172.
12 Stengers, *Making Sense in Common*, p. 95.
13 Latour, "A Textbook Case Revisited," p. 3. The "proof" of Latour's

relativism has been monotonously repeated by all those who seem to have read only one article by him (fitting on a single page): "Jusqu'où faut-il mener l'histoire des découvertes scientifiques?" ("How far should the history of scientific discoveries be taken?") first published in *La Recherche*, was reprinted in Bruno Latour, *Chroniques d'un amateur des sciences*, pp. 51–3. In 1976, the mummy of Ramses II arrived in Paris. *Paris Match* humorously captioned a photo: "Nos savants au secours de Ramsès II tombé malade 3 000 ans après sa mort." ("Our scientists come to help Ramses II who fell ill 3,000 years after his death.") Latour wonders if there is an anachronism in saying that the "Pharaoh died of tuberculosis discovered in 1882"? He wrote: "History writes its mark on the objects of science and not only on the ideas of those who discover them." This was enough to trigger the ire of the rationalists who finally had their proof! Later, in *Pandora's Hope*, Latour would say he had found the right way to talk about it: "After 1864, airborne germs were there all along" (p. 173). But reading a book in its entirety would be too much to ask of critics . . .

14 Latour, "A Textbook Case Revisited," p. 12.
15 Stengers, *Thinking with Whitehead*, p. 246.
16 Stengers, *Thinking with Whitehead*, p. 245.
17 Stengers, *Thinking with Whitehead*, p. 234. Translation modified.
18 Stengers, *Thinking with Whitehead*, p. 230.
19 Stengers, *Thinking with Whitehead*, p. 229.
20 Gilles Deleuze, *Negotiations*, trans. M. Joughin, New York: Columbia University Press, 1990, p. 128.
21 Félix Guattari, *The Three Ecologies*, trans. I. Pindar and P. Sutton, London: Bloomsbury, 2014.
22 Stengers, *Making Sense in Common*, pp. 154–5.
23 Stengers, *Making Sense in Common*, p. 156.
24 Pignarre and Stengers, *Capitalist Sorcery*, pp. 117–18.
25 Stengers, *Making Sense in Common*, p. 171.
26 Stengers and Despret, *Women Who Make a Fuss*, p. 87.
27 Stengers, *Thinking with Whitehead*, p. 489, citing Gilles Deleuze, *Difference and Repetition*, trans. P. Patton, New York: Columbia University Press, p. 249.
28 Latour, *After Lockdown*, p. 128.
29 Souriau, *The Different Modes of Existence*, p. 214.

Bibliography

Aït-Touati, Frédérique and Coccia, Emanuele (eds) (2021) *Le Cri de Gaïa: Penser la Terre avec Bruno Latour*. Paris: La Découverte.
Bachelard, Gaston (2000) *The Formation of the Scientific Mind*, trans. Mary McAllester Jones. Manchester: Clinamen Press.
Balibar, Étienne (2020) *Passions du concept. Épistémologie, théologie et politique, Écrits II*. Paris: La Découverte.
Benveniste, Émile (1971) *Problems in General Linguistics*, trans. M. E. Meek. Miami: University of Miami Press.
Bourdieu, Pierre (2004) *Science of Science and Reflexivity*, trans. R. Nice. Chicago, IL: University of Chicago Press.
Boutedja, Houria (2017) *Whites, Jews, and Us: Toward a Politic of Revolutionary Love*. Los Angeles, CA: Semiotext(e).
Callon, Michel (2021) *Markets in the Making: Rethinking Competition, Goods, and Innovation*, trans. O. Custer. Princeton, NJ: Princeton University Press.
Castree, Noel and Willems-Braun, Bruce (eds) (1998) *Remaking Reality: Nature at the Millennium*. London: Routledge.
Chertok, Léon and Stengers, Isabelle (1992) *A Critique of Psychoanalytic Reason*, trans. Martha Noel Evans. Stanford, CA: Stanford University Press.
Chertok, Léon, Gille, Didier, and Stengers, Isabelle (2020) *Une vie de combats: De l'antifascisme à l'hypnose*. Paris: La Découverte.

Bibliography

Debaise, Didier (2006) *Un Empirisme spéculatif: Lecture de Procès et Réalité de Whitehead*. Paris: Vrin.
Debaise, Didier (2011) "Alfred North Whitehead. Les sujets possessifs," in Didier Debaise, (ed.), *Philosophie des possessions*. Dijon: Les Presses du réel, pp. 233–51.
Debaise, Didier (2015) "Les âmes du monde," in Fleur Courtois-L'Heureux and Aline Wiame (eds), *Étienne Souriau: une ontologie de l'instauration*. Paris: Vrin.
Debaise, Didier (2015) *L'Appât des possibles: Reprise de Whitehead*. Dijon: Les Presses du réel.
Debaise, Didier and Stengers, Isabelle (eds) (2015) *Gestes spéculatifs*. Dijon: Les Presses du réel.
Deleuze, Gilles (1990) *Negotiations*, trans. M. Joughin. New York: Columbia University Press.
Deleuze, Gilles and Guattari, Félix (1987) *A Thousand Plateaus; Capitalism and Schizophrenia*, trans. B. Massumi. Minneapolis, MN: University of Minnesota Press.
Deleuze, Gilles and Guattari, Félix (1994) *What is Philosophy?* trans. H. Tomlinson and G. Burchell. New York: Columbia University Press.
Derrida, Jacques (1981) *Dissemination*, trans. B. Johnson. Chicago: Chicago University Press.
Despret, Vinciane (2020) *Quand le loup habitera avec l'agneau*. Paris: La Découverte.
Despret, Vinciane (2021) *The Dance of the Arabian Babbler: Birth of an Ethological Theory*, trans. J. Bussolini. Minneapolis, MN: Univocal.
de Vries, Gerard (2016) *Bruno Latour*. Cambridge: Polity.
de Vries, Gerard (2017) *Nature as Event: The Lure of the Possible*, trans. M. Halewood. Durham, NC: Duke University Press.
Fassin, Didier (1999) "L'ethnopsychiatrie et ses réseaux. L'influence qui grandit," *Genèses* 35: 146–71.
Flek, Ludwik (1979) *Genesis and Development of a Scientific Fact*, trans. F. Bradley and T. J. Trenn. Chicago, IL: University of Chicago Press.
Ginzburg, Carlo (1980) "Morelli, Freud and Sherlock Holmes: Clues and Scientific Method," *History Workshop Journal* 9(1): 5–36.
Ginzburg, Carlo (1989) *Clues, Myths, and the Historical Method*, trans. John and Anne C. Tedeschi. Baltimore, MD: Johns Hopkins University Press.

Guattari, Félix (2014) *The Three Ecologies*, trans. I. Pindar and P. Sutton. London: Bloomsbury.

Haraway, Donna (1988) "Situated Knowledges: The Science Question in Feminism and the Privilege of Partial Perspective," *Feminist Studies* 14(3): 575–99.

Haraway, Donna (2016) *Staying with the Trouble: Making Kin in the Chthulucene*. Durham, NC: Duke University Press.

Harper, Kyle (2017) *The Fate of Rome: Climate, Disease and the End of an Empire*. Princeton, NJ: Princeton University Press.

Hustak, Carla and Myers, Natasha (2012) "Involuntary Momentum: Affective Ecologies and the Sciences of Plant/Insect Encounters," *differences: A Journal of Feminist Cultural Studies* 23(3): 74–118.

James, William (1907) *Pragmatism*. New York: Longmans, Green, and Co.

Latour, Bruno (1983) "Comment redistribuer le Grand Partage?" *Revue de synthèse* 110 (April/June): 203–36.

Latour, Bruno (1987) *Science in Action: How to Follow Scientists and Engineers through Society*. Cambridge, MA: Harvard University Press.

Latour, Bruno (1988) *The Pasteurization of France*, trans. Alan Sheridan. Cambridge, MA: Harvard University Press.

Latour, Bruno (1993) *We Have Never Been Modern*, trans. C. Porter. Cambridge, MA: Harvard University Press.

Latour, Bruno (1994) *Pasteur: une science, un style, un siècle*. Paris: Perrin.

Latour, Bruno (1995) *Le Métier de chercheur. Regard d'un anthropologue*. Paris: INRA éditions.

Latour, Bruno (1996) "On Interobjectivity," *Mind, Culture, and Activity: An International Journal* 3(4): 228–45.

Latour, Bruno (1996) *Aramis, or the Love of Technology*, trans. C. Porter. Cambridge, MA: Harvard University Press.

Latour, Bruno (1996) "Do Scientific Objects Have a History? Pasteur and Whitehead in a Bath of Lactic Acid," trans. Lydia Davis, *Common Knowledge* 5(1) (Spring): 76–91.

Latour, Bruno (1999) "Factures/Fractures: From the Concept of Network to the Concept of Attachment," *RES: Anthropology and Aesthetics* 36 (Autumn): 20–31, 22.

Latour, Bruno (1999) *Pandora's Hope: Essays on the Reality of Science Studies*. Cambridge, MA: Harvard University Press.

Latour, Bruno (2001) *Le métier de chercheur: Regard d'un anthropologue*. Versailles: Éditions Quæ.

Latour, Bruno (2002) "Il ne faut pas qu'une science soit ouverte ou fermée," interview with Jean-Marc Lévy-Leblond, *Rue Descartes* 3(4): 66–81.

Latour, Bruno (2003) *Un monde pluriel mais commun*. La Tour-d'Aigues: Éditions de l'Aube.

Latour, Bruno (2004) *The Politics of Nature: How to Bring the Sciences into Democracy*, trans. Catherine Porter. Cambridge, MA: Harvard University Press.

Latour, Bruno (2004) "Whose Cosmos, Which Cosmopolitics? Comments on the Peace Terms of Ulrich Beck," *Common Knowledge* 10(3): 450–62.

Latour, Bruno (2005) "What is Given in Experience? A Review of Isabelle Stengers *Penser avec Whitehead*," *Boundary 2* 32(1) (Spring): 222–37.

Latour, Bruno (2005) *Reassembling the Social: An Introduction to Actor-Network-Theory*. Oxford: Oxford University Press.

Latour, Bruno (2006) "Des sujets récalcitrants," in *Chroniques d'un amateur de sciences*. Paris: Presses des Mines.

Latour, Bruno (2006) "Le travail de l'image ou l'intelligence savante redistribuée," in *Petites Leçons de sociologie des sciences*. Paris: La Découverte.

Latour, Bruno (2008) "The Netz-Works of Greek Deductions – A Review of Reviel Netz's *The Shaping of Deductions in Greek Mathematics*," *Social Studies of Science* 38(3): 444–59.

Latour, Bruno (2008) "A Textbook Case Revisited: Knowledge Mode of Existence," in E. Hackett, O. Amsterdamska, M. Lynch, and J. Wajcman (eds), *The Handbook of Science and Technology Studies*, 3rd edn. Cambridge, MA: MIT Press.

Latour, Bruno (2010) *Cogitamus: Six lettres sur les humanités scientifiques*. Les Empêcheurs de penser en rond series. Paris: La Découverte.

Latour, Bruno (2010) *The Making of Law: An Ethnography of the Conseil d'État*, trans. Marina Brilman and Alain Pottage. Cambridge: Polity.

Latour, Bruno (2010) *On the Modern Cult of the Factish Gods*. Durham, NC: Duke University Press.

Latour, Bruno (2011) "Reflections on Etienne Souriau's *Les différents modes d'existence*," trans. S. Muecke, in Graham

Harman, Levi Bryant, and Nick Srnicek (eds), *The Speculative Turn*. Melbourne: re.press, pp. 304–33.

Latour, Bruno (2012) with Godmer, Laurent, and Smadja, David, "The Work of Bruno Latour: Exegetical Political Thinking," *Raisons politiques* 47(3): 115–48.

Latour, Bruno (2013) "Biography of an Inquiry: On a Book about Modes of Existence," *Social Studies of Science* 43(2): 287–301.

Latour, Bruno (2013) *An Inquiry into Modes of Existence: An Anthropology of the Moderns*, trans. C. Porter. Cambridge, MA: Harvard University Press.

Latour, Bruno (2013) *Rejoicing: Or the Torments of Religious Speech*, trans. J. Rose. Cambridge: Polity.

Latour, Bruno (2014) "Pourquoi Péguy se répète-t-il? Péguy est-il illisible?" in Camille Riquier (ed.), *Charles Péguy*. Paris: Les Cahiers du Cerf.

Latour, Bruno (2015) "Différencier amis et ennemis à l'époque de l'Anthropocène," in Didier Debaise and Isabelle Stengers (eds), *Gestes spéculatifs*. Dijon: Les Presses du réel.

Latour, Bruno (2017) *Facing Gaia: Eight Lectures on the New Climatic Regime*, trans. C. Porter. Cambridge: Polity.

Latour, Bruno (2018) "Esquisse d'un parlement des choses," *Écologie et Politique* 56: 47–64.

Latour, Bruno (2019) "Troubles dans l'engendrement," Bruno Latour interviewed by Carolina Miranda. *Revue du Crieur No. 14*. La Découverte/Mediapart.

Latour, Bruno (2020) "Le Covid comme crash-test," interview with Thibaut Sardier, *Libération*, 13 May.

Latour, Bruno (2021) *After Lockdown*, trans. J. Rose. Cambridge: Polity.

Latour, Bruno and Lépinay, Vincent Antonin (2009) *The Science of Passionate Interests: An Introduction to Gabriel Tarde's Economic Anthropology*. Chicago, IL: Prickly Paradigm Press.

Latour, Bruno and Stengers, Isabelle (1997) "Du bon usage de l'ethnopsychiatrie," *Libération*, 21 January.

Latour, Bruno and Woolgar, Steve (1979) *Laboratory Life: The Social Construction of Scientific Facts*. Beverly Hills, CA: Sage.

Latour, Bruno and Woolgar, Steve (1986) *Laboratory Life: The Social Construction of Scientific Facts*, 2nd edn with a Postscript, "Laboratory Life: The Construction of Scientific Facts." Princeton, NJ: Princeton University Press.

Lévi-Strauss, Claude (1949) *Structural Anthropology*. New York: Doubleday.
Marx, Karl and Engels, Friedrich (1888) *The Communist Manifesto*. Project Gutenberg Ebook.
Mitchell, Timothy (2013) *Carbon Democracy: Political Power in the Age of Oil*. London: Verso.
Morizot, Baptiste (2020) *Manières d'être vivant*. Arles: Actes Sud.
Nathan, Tobie (2001) *Nous ne sommes pas seuls au monde*. Paris: Seuil.
Nathan, Tobie (2004) *Du Commerce avec les diables*. Paris: Seuil.
Nathan, Tobie (ed.) (2006) *La Guerre des psys: Manifeste pour une psychothérapie démocratique*. Paris: Seuil.
Nathan, Tobie and Stengers, Isabelle (1990) *L'Hypnose, blessure narcissique*. Les Empêcheurs de penser en rond series. Paris: La Découverte.
Nathan, Tobie, Stengers, Isabelle, and Gille, Didier (1990) *Mémoires d'un hérétique*. Paris: La Découverte.
Nathan, Tobie, Stengers, Isabelle, and Hounkpatin, Lucien (1997) *La Damnation de Freud* (a play in four acts). Les Empêcheurs de penser en rond series. Paris: La Découverte.
Pignarre, Philippe and Stengers, Isabelle (2011) *Capitalist Sorcery: Breaking the Spell*, trans. and ed. A. Goffey. Basingstoke: Palgrave Macmillan.
Prigogine, Ilya and Stengers, Isabelle (1984) *Order Out of Chaos: Man's New Dialogue with Nature*. London: Verso.
Prigogine, Ilya and Stengers, Isabelle (1988) *Entre le temps et l'éternité*. Paris: Fayard.
Riquier, Camille (2011) "Charles Péguy. Métaphysique de l'événement," in Didier Debaise (ed.), *Philosophie des possessions*, Dijon: Les Presses du réel, pp. 197–231.
Serres, Michel (ed.) (1993) *Éléments d'histoire des sciences*. Paris: Bordas.
Shapin, Steven and Schaffer, Simon (1985) *Leviathan and the Air-Pump: Hobbes, Boyle, and the Experimental Life*. Princeton, NJ: Princeton University Press.
Souriau, Étienne (2015 [1956]) *The Different Modes of Existence, followed by, Of the Mode of Existence of the Work to be Made*. Introduction by Isabelle Stengers and Bruno Latour, trans. Erik Beranek and Tim Howles. Minneapolis, MN: Univocal.

Stengers, Isabelle (ed.) (1987) *D'une science à l'autre: Les concepts nomades*. Paris: Seuil.
Stengers, Isabelle and Schlanger, Judith (1989) *Les Concepts scientifiques: Invention et pouvoir*. Paris: La Découverte.
Stengers, Isabelle (1992) *La Volonté de faire science: À propos de la psychanalyse*. Paris: Seuil.
Stengers, Isabelle (ed.) (1994) *L'Effet Whitehead*. Paris: Vrin.
Stengers, Isabelle (1998) "Entre collègues et amis," in Pierre Verstraeten and Isabelle Stengers (eds), *Gilles Deleuze*. Paris: Vrin.
Stengers, Isabelle (2000) "Discipline et interdiscipline: la philosophie de 'l'écologie des pratiques' interrogee." *Nature Sciences Sociétés* 8(3): 51–8.
Stengers, Isabelle (2000) *The Invention of Modern Science*, trans. D. Smith. Minneapolis, MN: University of Minnesota Press.
Stengers, Isabelle and Bensaude-Vincent, Bernadette (2001) *Histoire de la chimie*. Paris: LaDécouverte.
Stengers, Isabelle (2001) *La Guerre des sciences aura-t-elle lieu? Scientifiction*. Paris: Seuil.
Stengers, Isabelle (2002) "Cultures: guerre et paix. Une semaine à Cerisy," *Ethnopsy* 4 (April): 7–38.
Stengers, Isabelle (2002) "Une politique de l'hérésie." Interview in *Vacarme* 19 (Spring): 4–13.
Stengers, Isabelle (2006) "Un engagement pour le possible." *Cosmopolitiques* 1, *La nature n'est plus ce qu'elle était* (April).
Stengers, Isabelle (2006) *La Vierge et le Neutrino: Les scientifiques dans la tourmente*. Les Empêcheurs de penser en rond series. Paris: Seuil,
Stengers, Isabelle (2010) *Cosmopolitics I*, trans. R. Bononno. Minneapolis, MN: University of Minnesota Press.
Stengers, Isabelle (2011) *Cosmopolitics II*, trans. R. Bononno. Minneapolis, MN: University of Minnesota Press.
Stengers, Isabelle (2011) "William James. Naturalisme et pragmatisme au fil de la question de la possession," in Didier Debaise (ed.), *Philosophie des possessions*, Dijon: Les Presses du réel, pp. 35–69.
Stengers, Isabelle with Despret, Vinciane (2014) *Women Who Make a Fuss: The Unfaithful Daughters of Virginia Woolf*. Minneapolis, MN: University of Minnesota Press.
Stengers, Isabelle (2014) *Thinking with Whitehead: A Free and*

Wild Creation of Concepts, trans. M. Chase. Foreword by Bruno Latour. Cambridge, MA: Harvard University Press.

Stengers, Isabelle (2015) "Que vas-tu faire de moi?," in Fleur Courtois-L'Heureux and Aline Wiame, *Étienne Souriau: une ontologie de l'instauration*. Paris: Vrin, pp. 63–85.

Stengers, Isabelle (2015) *In Catastrophic Times: Resisting the Coming Barbarism*, trans. Andrew Goffey. London: Open Humanities Press.

Stengers, Isabelle (2017) "The Insistence of Possibles: Towards a Speculative Pragmatism," *Parse Journal* 7 (Autumn).

Stengers, Isabelle (2018) *Activer les possibles, dialogue avec Frédérique Dolphijn*. Noville-sur-Méhaigne: Éditions Esperluète.

Stengers, Isabelle (2018) "The Doctor and the Charlatan," in Tobie Nathan and Isabelle Stengers, *Doctors and Healers*, trans. S. Muecke. Cambridge: Polity.

Stengers, Isabelle (2018) *Another Science is Possible: A Manifesto for Slow Science*, trans. S. Muecke. Cambridge: Polity Press.

Stengers, Isabelle (2019) "Pourquoi le paysan argentin a raison de dire que le soja OGM est 'méchant'?" *Terrestres* 5 (6 May).

Stengers, Isabelle (2019) *Résister au désastre*. Marseille: Wildproject.

Stengers, Isabelle (2022) *The Virgin Mary and the Neutrino: Reality in Trouble*, trans. Andrew Goffey. Durham, NC: Duke University Press.

Stengers, Isabelle (2022) *Making Sense in Common: A Reading of Whitehead in Times of Collapse*, trans. T. Lamarre. Minneapolis, MN: University of Minnesota Press.

Tsing, Anna Lowenhaupt (2004) *Friction: An Ethnography of Global Connection*. Princeton, NJ: Princeton University Press.

van Dooren, Thom and Chrulew, Matthew (eds) (2022) *Kin: Thinking with Deborah Bird Rose*. Durham, NC: Duke University Press.

Whitehead, Alfred North (1978 [1929]) *Process and Reality: An Essay in Cosmology*, ed. D. R. Griffin and D. W. Sherburne. New York: Free Press.

Whitehead, Alfred North (2011 [1934]) *Nature and Life*. Cambridge: Cambridge University Press.

Whitehead, Alfred North (2016 [1919]) *The Concept of Nature*. Project Gutenberg Ebook.

Index

abstraction 42, 122, 142
  epistemological 7
  mode(s) of 3, 57, 62, 64,
    102–3, 135
actants 9, 15–16, 25, 26, 27, 30,
  36, 64, 114
active voice 7
activists 118
  distinction between militants
    and 124
Actor Network Theory (ANT)
  73, 74
agents/agency 31, 36, 74, 124
agora, Latour's 108, 110
alienation 23, 71, 123–4, 128
Althusser, Louis 35
Amazonian forest
  and Latour's article on
    empirical philosophy 41–6,
    62
anaphora 106
anthropology/anthropologists 14,
  19, 20, 67, 68, 72, 95
  symmetrical 81
anthropomorphism 77
anti-fetishism 77

baboons 46–7, 137
Bachelard, Gaston 20, 40, 52,
  129, 139
Badiou, Alain 18–19
Balibar, Étienne 18
Bastide, Françoise 8
Beck, Ulrich 94, 95
beetroot example 132, 134
bifurcation of nature 7, 101–4,
  135
  and Latour 71, 101, 114, 120
  and Stengers 101–2, 103, 120
  and Whitehead 101, 103, 105,
    141
big concepts 126–7
biology 33, 62
  evolutionary 138
black boxes 30–1, 35, 37–8
Bourdieu, Pierre 20
Bouteldja, Houria 9
Boyle, Robert 24, 53, 79

Calculemus 85, 88–9, 133
Callon, Michel 8, 132
Calmette, Albert 80
Canguilhem, Georges 18–19

Index

capitalism
  and Latour 15, 17, 123, 125–6
  and Marxists 123
  and Stengers 115, 122–3, 126–7
Carson, Rachel 119
Castro, Eduardo Viveiros de 67
chemistry 1, 40, 54
Chertok, Léon 2, 6, 34, 35, 39, 64, 65
Climatic Regime, New 13, 112, 125
coffee grinds, reading of 65
collectives, scientific 81
colonization 113–14
commerce 143
common sense 139
  and good sense 92
  and Latour 18, 92, 99
  and Stengers 54, 92
common world 4, 84, 123, 142
consciousness raising 128
Conseil d'Etat 97, 97–8
cosmopolitanism 95
cosmopolitical Parliament 86, 89, 91, 93–4, 95
cosmopolitics 86–8, 94, 95, 123
COVID 85, 114
critical sociology 20, 47–8, 70, 74
critical thinking/thought 69, 70, 72, 92, 102
critical zone 28, 135

Dalcq, Albert 138
Darwinian science 62, 63, 141
Debaise, Didier 4
Deleuze, Gilles 1, 9, 62, 107, 109–10, 142, 144
Deleuze, Gilles and Guattari, Félix 76
  A Thousand Plateaus 47, 128
  What is Philosophy? 107
Descartes, René 22
description 23, 45, 62, 114

Descola, Philippe 8
Despret, Vinciane 64, 137
Dewey, John 9, 91, 124
diplomacy/diplomats 89–91, 107–8, 116, 141–2
domains
  movement to modes of existence from 99
Double Click 3, 99, 106
DSM 68, 82
Duhem, Pierre 104
Dumas, Alexandre 8–9
Durkheim, Émile 49

ecological crises 86
ecological intelligence 117
"ecologies of practice" 39
ecology 10–11, 57, 79–96, 85–6, 112, 114, 124, 134
economy 137–8
  distinction between economics and 133
  knowledge 46, 58, 116–17, 119
  Latour on 132, 133–4, 135–6
  political 127
ecosophy 142
empiricism 41–8, 105
  and Latour 41–4, 107
empowerment techniques 129–30
engendering 134, 135, 136
epistemological abstraction 7
epistemology 18, 19, 39, 53, 99, 106
  Bachelardian 14
  Latour on 19, 20–2, 23, 29, 30–1
ethnography 18, 19, 47, 50
ethnopsychiatry/ethnopsychiatrists 65–6, 67–8, 72, 75, 76, 77, 94
ethology 13, 77, 138
evolution 38, 62, 113
  and involution 113
evolutionary sciences 63

experimental laboratories 32, 37, 39, 45, 46, 56–8, 62, 72, 76, 137

factish(es) 69–70, 72–4, 98, 105
facts
  and Latour 22, 27, 28, 69–70, 113
  and networks 49
fetishes/fetishism 69, 72, 77
field science/scientists 46–7, 63–4
fieldwork/fieldwork sciences 62, 119
  and Latour 22, 38, 62
  distinction between laboratory science and 38
fieldwork laboratory 43–4
Foucault, Michel
  "regimes of truth" 99
Freud, Sigmund 38, 77–8

Gaia 12, 82, 104, 111–36, 141
Galileo 53–4, 56, 61, 115
  inclined plane of 53
Garfinkel, Harold 8
generalized symmetry, Latour's principle of 24–5, 45
generative apparatuses 142–3
generative process 128
genetically modified organisms *see* GMOs
geo-social classes 134
Georges Devereux Center 67, 68, 69, 72, 98
Ginzburg, Carlo 12–13, 38, 44, 50, 61
  scientists as hunters essay 63
GMOs (genetically modified organisms) 59–60, 75, 82, 103, 119–20
good sense
  and common sense 92
Gould, Stephen Jay 102, 138
  Wonderful Life 63

Great Divide 15, 55, 71–2, 80, 94, and Bourdieu 20
  and Latour 6, 17–19, 22–3, 66, 141
  and Marxists 127–8
  and Stengers 66, 81–2, 86, 92, 103–4, 128, 140, 142
Greimas, Algirdas 8
growth hormone laboratory experiments 29, 43
Guattari, Félix 62, 107, 142
  The Three Ecologies 115
Guillemin, Professor 15, 43

Hajmirbaba, Soheil 130–1
Haraway, Donna 9, 11, 143–4
healers/healing 65, 72, 76–7, 78
  war between doctors and traditional 66
Hegelian dialectic 116
hesitations 5, 44
  as necessary part of researchers' work 44–5
  and Stengers 3, 4, 5, 14
hiatus 106, 135
Hobbes, Thomas 79
homeopaths 89–90
humanism 109
Hustak, Carla 63
hybrids 80
hypnosis
  and Freud 77
  and Stengers 33, 35, 64
hypnotizers/hypnotized 6

induction 6, 138, 144
inquirers 91
instauration 6, 7, 105, 106, 109
interdependence 113, 117, 118, 134
interdisciplinarity 44–5
involution 113, 133
IPCC (Intergovernmental Panel on Climate Change) 10, 112

Index

James, William 5, 9, 106

Kafka, Franz
  Metamorphosis 114
Kant, Immanuel 86, 102, 142–3
knowledge economy 46, 58, 116–17, 119
Kuhn, Thomas
  "paradigm" idea 54

laboratories 13, 26–7, 28, 43, 56–7
  bifurcation of nature in 102
  experimental 32, 37, 39, 45, 46, 56–8, 62, 72, 76, 137
  fieldwork 43–4
  Guillemin's 43
  and Latour 46, 49, 51, 57, 73, 74
  leaving the 82–3
  and nonhumans/actants 25, 36
  Prigogine's 55
  religious speech and work in 98–9
  and Stengers 36, 53, 54, 55–6, 58
  theoretical-experimental 64
Las Casas 95
Latour, Bruno
  and activists 124
  After Lockdown 15, 113–14, 129, 130
  and agora 108, 110
  Amazonian forest/Boa Vista research 41–6, 62
  Aramis 31–2
  and bifurcation of nature 71, 101, 114, 120
  and black boxes 30–1, 35
  and capitalism 15, 17, 123, 125–6
  Cogitamus 3, 95
  and common sense 18, 92, 99
  and common world 4, 84, 123
  and critical sociology 47–8, 70, 74
  de-epistemologizing the sciences 22–33, 39
  and diplomacy 107–8, 110, 116
  distancing from relativism 23–4, 30, 47, 114
  distancing from Science and Technology Studies (STS) 73
  distancing from sociology of science 47–8
  on distinction between primary and secondary qualities 71
  on distinction between science and technique/technologies 31–3
  and double-click communication 99
  Down to Earth 15, 113
  and ecology 10–11, 85–6, 112, 114, 124
  and economy and economics 132, 133–4, 135–6
  and empiricism 41–4, 107
  and engendering/subsistence 134
  on epistemology 19, 20–2, 23, 29, 30–1
  and ethnography 47–8, 50
  Facing Gaia 15, 112–13, 124
  and factishes 67–73, 74
  and facts 27, 28, 69–70, 113
  and field science/field work 22, 38, 46–7, 62
  and Gaia 111–13, 114, 124–5, 134
  generalized symmetry principle 24–5, 45
  graphics and diagrams punctuating books 4–5
  and the Great Divide 6, 17–19, 22–3, 66, 141

Latour, Bruno (*cont.*)
  Inquiry into Modes of Existence 3, 8, 32–3, 51, 69, 73, 74, 79, 90–1, 99, 104, 106, 107, 112, 114, 209
  Irreductions 9, 15–18, 20, 37, 70, 79, 87, 113, 135
  Jubiler ou les tourments de la parole religieuse 98
  and laboratories 46, 49, 51, 57, 73, 74
  Laboratory Life 14–15, 20–1, 26–7, 51, 97, 113
  The Making of Law 97–8
  and Marxists 9
  on materialism/materialists 20, 124, 131–2
  on modern world/Moderns 14, 15–17, 79–81, 112, 119, 124–5
  and modes of existence 67, 97–100, 104–5, 106, 106–7, 138
  and Nathan's ethnopsychiatry 67–9
  networks and termite example 27–8, 49–50
  On the Cult of the Factish Gods 67–70, 72–3
  opposition to Stengers' cosmopolitics 95
  and Parliament of Things 82–6
  on Péguy 8
  philosophical influences 9
  The Politics of Nature 2, 59, 75, 79, 83, 86, 111–12, 126, 132
  and rationalism 54
  Reassembling the Social 48, 71
  Rejoicing 5
  as a relationist 23, 114
  and relativists 23–4, 30, 54
  on religious speech 98–9
  reversibility of scientific inscriptions 28–9
  Science in Action 12, 18, 23–4, 34, 50–1
  and scientists 25, 47, 85, 107, 120–1, 136, 139
  on social classes 134–5
  and social sciences 64, 94, 98, 99
  and sociology 25–6, 49
  and sociology of science 49–51, 55
  "The Sphinx of the Work" (with Stengers) 100, 104–5
  on the symbolic 68
  taking nonhumans/actants into account 24–30, 36, 79–80, 81, 112, 114
  things in common with Stengers 4–5, 9
  and universality 87, 88
  We Have Never Been Modern 79–80, 83, 85
  and Whitehead 141
  writing style 3, 8
law
  and Latour 97–8
  "leap of the imagination" 5–6
Leibniz, Gottfried Wilhelm 3, 54, 88
Lévi-Strauss, Claude 65
Lovelock, James 111

McClintock, Barbara 138
Maniglier, Patrice 121
Margulis, Lynn 63, 119, 137
Marx, Karl 125, 127–8
Marxism/Marxists 20, 115, 122, 125
  and Stengers 9, 118, 123–4, 127, 128
materialism/materialists 20, 124, 131–2
mathematics 62

Index

mediations 99, 106
Method 37
"middle voice" 7–8, 102
migrants 66
militants
  distinction between activists and 124
model makers 63
models 61–2
modernization 119
Moderns 15–17, 79–81, 86, 87, 99, 107, 112, 119, 124–5, 140, 142
mode(s) of abstraction 3, 57, 62, 64, 102–3
mode(s) of existence 67, 92–110, 134, 138
musical metaphor 6
Myers, Natasha 63

Nathan, Tobie 2, 39, 65–9, 75, 76, 95
  "Dealing with Devils" 77
natural selection 62
nature, bifurcation of see bifurcation of nature
neo-pagan witches 9, 72
neonicotinoids
  reauthorization of by French government 132–3
networks 45, 49–50, 85, 95–6, 97
  and centers of calculation 50
  construction of standardized 50
  and facts 49
  scientific 50, 59
  termite example 27–8
neurosciences 37
nonhumans 15, 24–30, 36, 79–80, 81, 85, 88, 95, 112, 114, 132

objectivity 29, 31, 39, 40, 54, 93, 107, 117, 141
obligations 35–6, 38, 39, 58–9, 65–6, 90, 93, 115, 117, 129, 130
opportunism 128, 139

palavers 87
Parliament, cosmopolitical 86, 89, 91, 93–4, 95
Parliament of Things 79–96
passive voice 7–8
Pasteur, Louis 22, 24–5, 29, 30, 36, 54, 61, 72–3, 74, 80, 97, 114
Pasteur–Pouchet controversy 22
Péguy, Charles 8, 14
pharmaceutical industry 139
physics 34, 35, 53, 104
  meta 24, 107
  relations between chemistry and 51, 54
Pignarre, Philippe and Stengers, Isabelle
  Capitalist Sorcery 126–7
placebo effect 66, 90
Plotinus 6
Poincaré, Henri 104
politics 139–40
  cosmopolitics 86–8, 94, 95, 123
  and science in the Parliament of Things 82, 84
politics of science
  and Stengers 52–60
Popper, Karl 52
Pouchet, Félix 22
power
  Stengers' characterization of science by threefold 56, 61, 64
Prigogine, Ilya 2, 34–5, 55, 101
Progress 58, 103, 107, 115, 118, 129, 139, 140
property 134
pseudo-sciences
  distinction between sciences and 37

psychiatrists 8, 67
psychism 98
psychoanalysis 34–5, 68, 77–8
 Stengers' critique of 31, 34, 35, 37, 38–9, 75

rationalists/rationality 37, 53, 55, 74
 and Latour 54
Reason 14, 39, 140
recalcitrance 3, 59, 64, 127
relationists 23, 114
relativism/relativists 20, 23–4, 30, 47, 54, 74, 81, 107, 114
relays 122, 128–9
reliable witness 24, 75, 79, 82, 90
religious speech 98–9
representations 29, 67–8, 80
requirements 58–9, 61, 115, 117
restricted symmetry principle 23

Sauvons la recherche (save research) movement 117
science and technology studies (STS) 19, 25, 26, 30, 73, 114
science wars 10, 52
scientific collectives 81
scientific knowledge 18–19
scientific spirit 129
scientists 19, 28, 29, 35, 116–18, 129
 dependence on private interests 116–17
 discursive regimes used by 58
 "field" 63
 financing of their research 118
 joy of 91, 137
 and knowledge economy 116
 and Latour 25, 36, 39, 47, 85, 107, 120–1
 obligations of 117
 and Parliament of Things 93

 and Stengers 40, 47, 52, 53, 54, 55–7, 59, 64, 82, 103, 117–18
 vulnerability of 117–18
Serres, Michel 9
Shapin, Steven and Schaffer, Simon
 Leviathan and the Air-Pump 24, 79
simians, sociology of 26
situated knowledge 58
social classes 134–5
social constructivism 10, 25, 30, 43, 70, 73, 74
social sciences 22, 64, 71, 99
society–nature divide *see* Great Divide
sociology 25–6
 critical 20, 47–8, 70, 74
sociology of science 47–8, 54–5
 and Latour 49–51, 55
Socrates 140
Sokal and Bricmont affair 10, 52
Souriau, Étienne 7, 105–6, 108–9
 Deleuze and 109–10
 The Different Modes of Existence 100, 144–5
speculation 91–2
speculative philosophy 138, 142
speech impediments 4, 102, 141, 144
Spinoza 83
 Tractatus Theologico-Politicus 15
Starhawk 9, 72, 129
Stengers, Isabelle
 and alienation theory 71, 123–4
 Au temps des catastrophes 2, 111
 and bifurcation of nature 101–2, 103, 120
 and big concepts 126–7
 and black boxes 34–5, 38
 and capitalism 115, 122–3, 126–7

Index

Capitalist Sorcery (with Pignarre) 126–7
as a chemist 54
and Chertok 6, 34, 35, 39, 64, 65
and commerce 143
and common sense 54, 92
and cosmopolitical Parliament 85–6, 89, 91, 93–4, 95
Cosmopolitics 74, 85
and cosmopolitics 86–8, 94, 95, 123
on critical sociology 70
critique of psychoanalysis 31, 34, 35, 37, 38–9, 75
and Darwinian science 62, 63
definition of the scientific concept 35
and diplomats/diplomacy 88–91, 142
and disamalgamating the sciences 34–40
on distinction between science and technology 32–3
on distinction between sciences and pseudo-sciences 36–7
distinction between unveiling and characterizing 71
and Dumas 8–9
and ecology 57, 85, 93
and factishes 73–4
and Gaia 118–19, 122
and generative apparatuses 142–3
and genetically modified organisms (GMOs) 59–60, 119–20
and the Great Divide 66, 81–2, 86, 92, 103–4, 128, 140, 142
and hesitations 3, 4, 5, 14
and hypnosis 33, 35, 64
Hypnosis, Between Magic and Science 75–8

and induction 6, 138, 144
and inquirers 91
The Invention of Modern Science 2, 9, 12, 52–7
and involution 113
and laboratories 36, 53, 54, 55–6, 58
and "leap of the imagination" 5–6
and Marx/Marxists 9, 118, 123–4, 127, 128
and the "middle voice" 7–8
and models 61–2
and Nathan 65–6
and neo-pagan witches 72, 129
and networks 85
and obligations 35–6, 38, 39, 58–9, 65–6, 115, 117, 130
Order Out of Chaos 34
philosophical influences 9
and "politics of science" 52–60
Réactiver le sens commun 7–8
"Scientific Black Boxes" lecture (1989) 34–5
and Souriau 108–9
and speculation 81–2
"The Sphinx of the Work" (with Latour) 100, 104–5
and techniques 74–6, 77
and theoretico-experimental sciences 53, 54
things in common with Latour 4–5, 9
Thinking with Whitehead 5, 7, 8, 16, 33, 100–2, 103–4, 107
threefold power of science 56, 61, 64
on transplantation 57
and utopianism 124
La Vierge et le Neutrino 53, 117
and Whitehead 75, 100–3, 141
writing style 3, 8

string games 143–4
Strum, Shirley 26, 46, 137
symbolic 68
symmetrical anthropology 81
symmetry
  Latour's principle of generalized 24–5, 45

Tarde, Gabriel 49, 135
technique(s)
  distinction between science and 31–2
  and Latour 75
  and Stengers 74, 75–6
technology
  distinction between science and 31–3
termites 28
theoretical-experimental sciences 36, 53, 54, 61, 62, 63, 64, 137
theories
  and data 51
Tournier, Michel
  Friday – Or the Other Island 15
trance 76
transplantation 57
Tsing, Anna
  Friction 128

unconscious
  psychoanalytic 78
  universal 68
universal/universality 21, 50, 80, 87–8, 112
universities 116
unveiling
  distinction between characterizing and 71
utopia/utopianism 87, 124, 125

Valladolid controversy 94–5
Van Dyck, Barbara 117
vigilance 102–3
voice, middle 7–8

Whitehead, Alfred North 2, 9, 24, 36, 75, 100–3, 138
  and bifurcation of nature 101, 103, 105, 141
  "philosophy of organism" 111
  and Souriau 7
  "trick of evil" 115
  see also Stengers, Isabelle: *Thinking with Whitehead*
witches 39, 129–30
  neo-pagan 72
wolf and three little pigs story, the 93
Woolgar, Steve 14